Future of Business and Finance

The Future of Business and Finance book series features professional works aimed at defining, describing and charting the future trends in these fields. The focus is mainly on strategic directions, technological advances, challenges and solutions which may affect the way we do business tomorrow, including the future of sustainability and governance practices. Mainly written by practitioners, consultants and academic thinkers, the books are intended to spark and inform further discussions and developments.

More information about this series at http://link.springer.com/series/16360

Lex A. van Gunsteren · Arnold G. Vlas

The License Giver Business Concept of Technological Innovation

A Game of Excellence

 Springer

Lex A. van Gunsteren
Noordwijk, The Netherlands

Arnold G. Vlas
Leiden, The Netherlands

ISSN 2662-2467   ISSN 2662-2475   (electronic)
Future of Business and Finance
ISBN 978-3-030-91122-5   ISBN 978-3-030-91123-2   (eBook)
https://doi.org/10.1007/978-3-030-91123-2

This Springer imprint is published by the registered company Springer Nature Switzerland AG
The registered company address is: Gewerbestrasse 11, 6330 Cham, Switzerland

Foreword

This short book summarises a life time of experiences of two practitioners who were deeply involved in technological innovation in large industrial enterprises, and who took on the roles of researcher, developer, engineer, patent holder, planner, business man, teacher, and consultant to companies and governments. It is a story told from the inside out—from those who were standing with two feet in the middle of practice, as opposed to those who tend to provide an outside-in view. Standing knee-deep in the swamp, they had to address uncertainties and complexities in real time, in contrast to those who stand on the slopes of the hill, surrounding the swamp, and who, with the benefit of hindsight, tease out stories and lessons as to what happened.[1] The authors state upfront that the view from the swamp is very different from the view from the hill—'two separate worlds'. In particular, following Schön (1971)[2] they take issue with those researchers who construe stories by imposing a retrospective and rational view of what happened—'a myth'—and subsequently derive unhelpful conclusions and recommendations from such misguided undertakings. The book tries to remedy this issue by not only telling what happens, but how and why it went wrong at times, and above all about 'how to think and what to do better' in the process of technological innovation by providing a fairly comprehensive map. The book is geared in particular towards the most senior executives in industrial firms as well as politicians, who take make-or-break decisions in funding and using technological innovations, and yet often have 'superficial knowledge' about what it takes to manage technological innovation.[3] It is also instructive for senior executives (e.g. 'how to establish a license fee?') and a cautionary tale for those deeply involved in actual technological innovation in large firms (e.g. the danger of overrating technical merits). At the very centre of the authors' quest is how firms can use high-quality technological designs ('inventions') that are functionally superior compared to existing designs, that can help society, and generate above-average profits ('innovations'). Excellence in design is defined as a combination of 'fitness for purpose', 'technologically balanced', and using 'state-of-the-art technologies'.

[1] See D. A. Schön, The reflective practitioner: how professionals think in action, 1984.

[2] See D. A. Schön, Beyond the stable state, 1971; D.A. Schön, Technology and change, 1967.

[3] The book distinguishes itself from practitioner accounts which are either 'tell-tale books' or provide 'a rosy picture of the actions' of the protagonist.

In general, a firm can either develop such design leadership in-house or use an externally developed design. In the former case, the authors warn that trying to hold on to the exclusive use of the patented invention for in-house use, or sales exclusively through its own commercial operation, will lead others to become potential adversaries. These others may then either try to beat the patent with a better one (if possible) or circumvent it. This is where the inventor-firm becomes 'a license giver' to others who become 'license takers'. The focus of the book is on managing the business of technological innovation, based on the concept of license giving, and how to implement the concept, in particular to assure high-quality, excellence, that is at the centre of competitive advantage once a license taker adopts the innovation.

The book gives many examples of technological innovations that establish design leadership, in particular in the shipbuilding industry, and in the computer industry.

The former is one of the very first industries right after WWII, where a 'vertical' industry structure (a few large quasi-monopolistic firms have their own exclusive in-house suppliers) is gradually replaced by a 'horizontal' one (where smaller and more nimble firms, buy from outside supplier, who service many firms). As such, the shipbuilding industry contains important lessons for similar industrial transformations in, for example, the computer industry after 1980.

A vertical structure, designed around 'an artefact'—like a ship with a steel hull, a combustion-engine car, a telephone service, or a mainframe computer—begins to buckle when its in-house suppliers (part of the vertical structure) can no longer keep up innovating for a smaller captive market. Outside part suppliers, servicing many quasi-monopolistic 'verticals', are no longer bound by their internal constraints. With larger markets for their products and services, they can invest in radical innovations, design excellence and leadership, establish a dominant design, reap the benefits, and continue to create new generations. The manufacturing and innovation for such parts is then outsourced by 'verticals', and over time offshored. Microsoft and Intel are good examples, creating conditions for firms like Dell to emerge (discussed as a case).

With the emergence of Japanese shipbuilders after WWII, such industrial transformation took place early on in the shipbuilding industry, at times run by legendary captains of industry. Propeller design was an early product to be supplied from the outside. Propeller design, in combination with the steering of a ship, is a distinct and very critical element, much like a CPU in a computer, which determines speed, cost, and manoeuvrability of ships. Ship propulsion can be well understood through mathematical modelling, tests in towing tanks, the cavitation tunnel, the vacuum tank, and ultimately in practice.

The book is replete with examples where 'non-conventional designs' are not even well managed in the propeller manufacturing firm itself, let alone the ship designers and ship yards, or the client who ordered the vessel. Such a complex and dynamic business system becomes even more complicated with competitive pressures on 'national champions', leading to mergers, consolidation, and rationalisation, not infrequently with government support, funding, and intervention. In such circumstances, 'non-conventional' but excellent designs ('inventions') face a very challenging trajectory to become commercially accepted ('innovations') even though they are critical to industrial renewal. The book even cites an example of a firm accepting an unconventional design, but then after an unexpected failure falling back in a kind of kneejerk reaction on a conventional design.

In these complex, dynamic, and turbulent industrial settings, the authors frame questions like: What does it take to convert design leadership (patented inventions) into commercial use (technological innovation)? How to manage a process that recognises the commercial potential of a scientific and engineering invention and incorporates it into a larger system's design for the benefit of the firms involved and society? Such a process may span the boundaries of many departments within an organisation, and across many organisations, including large firms, government agencies, and politicians.

What makes the process particularly complex and challenging is that innovation often takes place in an organisational setting that is still reaping the benefits of a previous innovation, around which structures, systems, processes, as well as values, norms, practices, and professional and organisational identities, careers, and power have shaped themselves, by design. Moreover, this process of technological innovation has to be managed repeatedly over a long period of time, leading from one generation of offerings to another to retain design leadership. This process is framed by the authors as 'a game of excellence', in which 'only one can be the best', like in an Olympic sports contest.

It is important to be mindful of the limits to this metaphor. A sports contest may take place in a very brief episode of time, with all credible competitors present in the same arena at the same time. The winner is a winner on that particular day, and a race a week later may lead to a different outcome, even though the winner becomes the 'reigning Olympic champion'. Most critically, in sport, 'the rules of the game' are set 'in stone', known long in advance, and meticulously followed during the contest.[4] More radical technological innovation intends to 'change the rules of the games as the game is being played² (things become possible that were not possible before). Moreover, the rules are affected by larger system issues. For instance, a more efficient propeller design may suddenly look less interesting after a steep drop in fuel price, or the market for transportation may change due to changing demand of products to be transported. An illustrative example is the contest between Betamax (Sony) and VHS (Matsushita-Panasonic). The former was by all accounts a superior design, yet the latter won out when it became clear over time that consumers preferred longer tapes. After a hesitation by Sony, initial differences in production volume gave Matsushita a chance to quickly drive down cost and price and become in a matter of years the dominant design, outcompeting an initially technically superior design. Regardless, in this process of technological innovation, many

---

[4] In ship propulsion, the technical variables and constrains of flow can be modelled mathematically with great accuracy, and testing in towing tanks and cavitation tunnels, much like airplane propellers and wing design can be well understood through wind tunnels. Hence, for ship propulsion the sheer technical rules of the game in that sense are well understood. Even there, the actual design of propellers and additions to it are not only engineering problems but also manufacturing problems.

things that can go wrong, will go wrong if not properly attended to, and even then success cannot be guaranteed.

A good part of the book addresses issues of implementation of the license giver concept in the process of technological innovation in which, drawing on the work of Mauk Mulder,[5] a range of actors hold different positions of formal and informal power. Some engaged in the process of research, development, and engineering have deep insight into the specifics of the process of a particular technological innovation but lack the specific formal power to implement processes to foster the innovation; CEOs and politicians who do have specific formal powers to shape the organisation(s) may lack a deeper insight into the particulars of process of innovation itself. The authors make recommendations as to how to try to think about collaboration between the various actors in innovation through writing this book. Formal power of politicians and CEOs, concentrated in the top and parcelled out layer by layer,[6] is sometimes insufficient for successful implementation. A poignant critique of that view ('the president as clerk') is contained in Neustadt (1960)[7] who states that nothing might happen if the president issues an order but does not understand the political interests at stake. This view is supplemented by Allison (1971)[8] discussing (a) the limits of rationality in terms of available information on the case and on options, and (b) the power of standard operation routines once set in motion. The authors point to the severe limits placed on CEOs and politicians, and the necessity to understand and operate in an (inter)-organisation world of pluralistic interests (see also Ansoff 1979).[9,10] Crozier reported how top management of a tobacco firm was near powerless in the face of 'the strategic game' of the maintenance workers who were responsible for addressing breakdowns in the manufacturing equipment. In this ultra-rationally designed organisation, intended to eliminate all uncertainty, they cornered the one remaining source of uncertainty and monopolised the knowledge for addressing it by insisting that they, and they alone, could fix machines, and in order to prevent production workers from making repairs, all

---

[5] See Mauk Mulder, The daily power game, 1977, and The Logic of Power...Alliances, 2012. As corporate planners of the Royal Boskalis Westminster Group, Lex van Gunsteren, and I worked extensively with Mulder in the early 1980, among others, on the functioning of the Executive Committee after divisional general managers had become members of the committee. Contrary to rational expectation, rather than facilitating top management decision-making, it made it more problematic. Mauk Mulder, 'The Logic of Power', 2010, http://www.maukmulder.nl/download.php

[6] For the most sophisticated view beyond the most tradition view of the rational organisation, see C. I. Barnard, Functions of the executive, 1938, and building on that, H. A. Simon, Administrative behavior, 1947.

[7] R. Neustadt, Presidential power and the modern presidents: the politics of leadership, 1960.

[8] G. Allison, Essence of Decision: Explaining the Cuban Missile Crisis 1971.

[9] H. I. Ansoff, Strategic management, 1979.

[10] While at the Royal Boskalis Westminster Group, I attended a seminar of Ansoff in Brussels (1980) were he discussed in detail the limits of the rational organisation by review Robert McNamara's attempt to transform the Department of Defence in the early 1960s, only to conclude that as soon as he left everything reverted to the way it was—as if McNamara had changed the straight line of a rubber band by pinning it down elsewhere only to see the rubber band return to its original position once the pin (McNamara) was removed.

documentation about the machines was carefully stacked away. In doing so, they held a powerful and indispensable grip on the organisation.[11] Bower found that in large complex industrial organisations (e.g. a chemical manufacturer) depending on highly specialised technical people, top management was faced with investment proposals from deep down in the organisation that 'trickled up', posing 'faits accomplis' that they could only rubberstamp.[12] All this suggests that those engaged in the process of technological innovation *actually* have significant informal power. Moreover, some CEOs, perhaps with CEO as Andy Grove and Steve Jobs as good examples, have themselves intimate knowledge of the process of technological innovation. So, while CEOs and politicians have formal power, they have to engage productively the expert insight of scientists and engineers to make the process work. The latter have indispensable, non-trivial informal power.

The book is replete with advice, systematically laid out. The authors start with an inventor (individual or an organisation) with a superior technological design that has been patented and could be licensed. This design leadership is an accomplishment all in itself, that few firms ever achieve (differentiating themselves through technological innovation), and that for many firms is beyond their reach and that they should not aim for it. In a horizontally tilting industrial structure, the holder of the patent should not only necessarily build off the patent themselves and refuse to license—i.e. to keep it proprietary, and exclusive as if it were a vertical structure. Nor should other firms, in need of strategic renewal, try to develop it themselves but rather license the design invented elsewhere (imitate technological innovation and differentiate on other aspects).

This sets up the relationship and the dynamic of 'license giver' and 'license taker'. They distinguish between four business identities—deeply embedded norms and values in the organisation: 'The license giver' which developed a design and the specifications for a replicable product, the 'license taker' which appropriates the design, and then two business identities ('consultant' and 'jobber') which operate in a fundamentally different way, in that they provide 'capacity' for working on given specifications provided by a client. It is critical to keep these distinctions in mind, and the authors go as far as to advice not to mix these identities as they represent incompatible values, thus leaving an organisation 'stuck-in-the-middle', an identity crisis of sorts.

Overall, these distinctions are a useful reminder of different strands of practice, and the importance of awareness of those differences to diagnose garbled communication, misunderstanding, tensions, and conflict in management. There is a long history of research in management,[13] called 'structural contingency theory', in response to the idea of 'one best way', that holds that different situations and environments require different organisational structures, extended to 'contingent

---

[11] M. Crozier, The bureaucratic phenomenon, 1964.

[12] J. Bower, Managing the resource allocation process, 1970.

[13] Tom Burns and G.M. Stalker, The management of innovation, 1961; P. R. Lawrence and J. W. Lorsch, Differentiation and integration in complex organizations, 1967; J. D. Thompson, Organizations in action, 1967.

leadership styles'.[14] While creating separate organisation units may make sense, nowadays, many organisations may hold both types in one firm, for example, a real estate development firm that had its own construction firm which also operated for other clients.[15] Increasingly, 'product' firms also deliver 'services' to clients, as an integrated offering.[16] Strategic agility (different from flexibility) may require managers to be able to keep both models in mind at the same time without going crazy, called by the poet Keats (1817) 'negative capability'.[17] The authors acknowledge the need for such capability in management, when they indicate that it is possible to transform a business organisation from one of the key identities to a license giver.

Next they go through the process of innovation step by step with ample illustrations. They begin by demonstrating that generating non-conventional technical inventions is difficult in the vertically integrated, top-down industry structures by tracing the multiple spin-offs that eventually led to the founding of Intel. In those 'vertical' structures, they warn, loyalty of engineers is to their professional organisations and technology leaders, not the organisation. To obtain their commitment, a loosely structured form of organisation, called adhocracy, is required. Next they identify 'the fear of innovation' in organisations resulting from a perceived threat to the dominant coalition in power, and the fending off of uncertainty, inherent in technological innovation. Uncertainty pertains to the lack of understanding of causality when a non-conventional invention is introduced into a system. This is very different from risk, whereby causality is understood but the outcomes are subject to probability. Uncertainty often leads to anxiety or anguish, a feeling often intuitively avoided by managers.[18] Even great persistence and creativity can thwart the best efforts to introduce unconventional designs, and several examples are provided to illustrate this.

The degree to which the introduction of an invention can be stymied is illustrated with the classic case of a very perceptive British officer in the Far East (Morrison 1966),[19] who noticed how a gunner with a stationary cannon on deck was compensating for the roll of the ship in the waves, improving his accuracy (which was

---

[14] See e.g. F. E. Fiedler, A Theory of Leadership Effectiveness, 1967; R. R. Blake and J. Mouton, The Managerial Grid: The Key to Leadership Excellence, 1964.

[15] See W. J. A. M. Overmeer, Corporate inquiry and strategic learning: the organizing major strategic change: the role of surprise and improvisation in organizing major strategic change, MIT doctoral thesis, 1989.

[16] R. Normann and R. Ramirez, Designing interactive strategy, 1995.

[17] 'Negative Capability, that is, when a man is capable of being in uncertainties, mysteries, doubts, without any irritable reaching after fact and reason' J. Keats, The Complete Poetical Works and Letters of John Keats, 1899.

[18] C. Argyris and D. A. Schön, Theory in practice: increasing professional effectiveness, 1974.

[19] E. Morison, Men, machines and modern times, 1966, Ch. 2. 'Gunfire at sea'.
     In this case, the British officer was the inventor, the US Navy officer was the Product Champion who fought to the bitter end for the acceptance of the invention. These two roles are often played by the same person, for instance, Chester Carlson who took his Xerox invention too well over 20 companies before he was able to make a sale. No one except Carlson himself envisioned the need for this process or believed it was practical.

known to be terrible, throughout). A US navy officer picked up the idea and experimented, showing significant improvement in accuracy. The invention was rejected over and over again, and when finally experiments were conducted in the USA, the experimental setting did not simulate the roll of the ship, thus showing no improvements over the conventional ways. It was only when the officer wrote to the President of the USA, Roosevelt, who had been in charge of the Navy, that he finally 'got an ear' and was subsequently appointed to 'teach the Navy how to shoot'. Even then, the officer ran into massive resistance. It is another striking example of how an organisation can fend off non-conventional inventions, with demonstrated dramatic increases in performance.

The organisational dynamics that make this happen were coined 'dynamic conservatism' by Schön (1971).[20] He concluded that organisations will try everything to remain stable, as its managers have come to live in a world in which they are successful (i.e. competent), and know how to be successful—they know the organisational causality. A challenge to that world would require a rethinking of existing norms and values, and development of new ones, including possibility of new professional identities. In the process, managers and engineers will encounter situations in which what they are no longer necessarily successful at first. This can be embarrassing and threatening to the point that managers intuitively try to control the situation, often unilaterally, thus impeding their own and others' learning of new relevant skills.[21]

It is not only the license taking organisation that can thwart the process, the inventor, the inventing organisation, can play a role as well. The authors mention that the license giver may overrate the technical merits of the invention while underrating the effort of the license taker to commercialise the invention. In addition, the license taker organisation may suffer from a 'not-invented-here syndrome' and be unwilling to embrace an idea from the outside that the organisation had not thought about, could not develop, or is reluctant to pay the fee for. This could well be more pronounced in an organisation that was once an innovator but that had curtailed its innovative capability by focusing increasingly on harvesting from that innovation and eliminating 'extraneous' efforts, thus creating a well-oiled machine at first, and a bureaucracy over time, unable to develop unconventional innovations.

These are the type of dynamics the authors describe and illustrate. Add to this that a patent is merely a legal document, and it becomes clear that the transfer of knowledge underlying the patent, and the process of technological innovation and strategic renewal is fraught with difficulties that can stop the process in its tracks, or lead to costly delays, or missing the window of opportunity. Is it any wonder that in

---

[20] D. A. Schön, Beyond the stable state, 1971 Defending the stable state is different from the idea of 'protecting the status quo' or the observation that 'the more things change, the more they stay the same'. A great deal of energy and action is mobilised, giving the appearance of dynamism and addressing the problem, while in fact the problem(s) are only addressed to the extent that they can be addressed with an existing repertoire of action, within a stable frame.

[21] C. Argyris and D. A. Schön, Theory in practice: increasing professional effectiveness, 1974.

the end the authors lament that excellence is not in good currency, and that attempts to elevate awareness of excellence are considered a taboo?

Nevertheless, the authors provide a concrete way out this labyrinth of hurdles and obstacles—a practitioner's map grounded in appreciation of the entire trajectory of technological innovation (the system). It involves an understanding of business strategy for exploiting the innovation as an issue of organisational identity, and implementing that strategy by understanding the use of power, which in turn is determined to a large extent in understanding the specifics of the technology and the specifics of the economics, in their situational context.

Excellence in design of the technology has to be accompanied by excellence in mapping the process towards successful innovation, and engaging and educating senior executives, CEOs, as well as heads of critical agencies, and politicians, with realistic assessments and projections the technological promise and its economic potential.

If senior executives and politicians are not versed in the technology and its economics, engineers have to understand the power dynamics, the politics, the social interactions, and the defensiveness non-conventional inventions may generate.

This is somewhat counterintuitive for scientists, developers, and engineers who often seem to move into those professions in order to avoid the social, organisational, and political aspects. But the authors show that it is critical for success, if that is what they want. Persistence, stamina, and creativity are not enough.

Willem J. Overmeer
Weston, MA, USA

Preface

The word *innovation* surfaces frequently in the media, in political campaigns, in governmental policy statements, in annual reports of corporations as well as in everyday social conversations. Once used to describe the process of making an invention commercially viable, the word has become overused, inducing us to notice:

1. More often than not, the espoused statements on innovation are made by people who themselves have never been deeply involved in innovating endeavours, nor have studied in-depth the complex process of technological innovation.
2. Their views are based on their seemingly unshakable belief in the *rational view of innovation*, the erroneous assumption that innovation is *manageable*, a technocratic viewpoint clearly in conflict with our experiences in the maritime and computer industries.
3. This combination of lack of knowledge of innovation and the belief that it can be rationally managed is a recipe for unfulfilled promises from technological innovation.

This state of affairs of an erosion of meaning leading to widely prevailing misconceptions shows striking similarity with how Operational Research (OR) evolved in the last half century. Its development goes back to the Second World War in the UK, when OR resolved the crucial question: How should the limited number of available radar stations be located to attain the highest probability of timely intercepting Göring's Luftwaffe bomber airplanes? The success of OR in the war led to widespread use beyond the military, eventually gaining acceptance in academic, scientific, and managerial circles. Over time, though, this gain was accompanied by a loss of pioneering spirit, its sense of mission, and its innovativeness. By the mid-1960s, academics who had never practised OR in complex real-world situations taught most OR courses in American universities. Enthralled by the beauty and simplicity of the mathematics, they developed an obsession with techniques that were easy to teach. The original interdisciplinarity of OR completely disappeared. Nowadays, OR is referred to as a discipline, encapsulated in math departments. As a result of an unrelenting focus on techniques, problems suited for application of those particular techniques received increasing priority over addressing more relevant problems that did not need them. An exasperated Ackoff, who stood at the birth of OR, finally proclaimed in 1991 that 'the future of OR is past'.

The history of technological innovation shows a similar pattern. Innovation, initially the accomplishment of converting an invention into a commercial success, became 'an idea in good currency' in the rebuilding of nations right after WWII. Over time, public figures, politicians, and CEOs of large corporations, wanting to share in its success, jumped on the bandwagon. Innovation, too, became an academic field for research. 'Innovation' became packaged as if it could be introduced in a particular corner of the organisation where it would fit in and could be managed through an orderly process 'if management would only do X, Y, and Z...'. An enduring public debate regarding the importance of innovation evolved in which most of its participants had only superficial knowledge of what innovation really entails in actual practice. The result has been that misconceptions on technological innovation could get widespread acceptance.

Does that matter?

Yes! It does matter. Politicians, civil servants, and CEOs have more power than others. How they use their power is highly determinant of the success or failure of innovative undertakings. When they adhere to the misconceptions mentioned before, they will repeat the blunders that were made in the past over and over again in the future.

The purpose of our book is to enable to upgrade insight into the dynamics of technological innovation by offering our theories, concepts, and experiences.

A second purpose of our book is to convey our views to our students and participants of our courses.

We thank Willem Overmeer for his critical review.

Finally, we wish the reader to enjoy reading, like we have experienced great pleasure in writing during our COVID-19 confinement.

Noordwijk, The Netherlands   Lex A. van Gunsteren
Leiden, The Netherlands   Arnold G. Vlas
2022

Contents

List of Figures

List of Tables

About the Authors

Lex A. van Gunsteren (1938) is a business consultant, lecturer, and innovator in marine propulsion. He graduated as a naval architect and received his PhD from Delft University of Technology, where in 1981 he was appointed as Professor in Management of Technology. He was, as professor of the Erasmus University Rotterdam, one of the pioneers of the Rotterdam School of Management, rated in the 1980s as the number one of the top-ten European business schools. After his military service as an officer in the ship design unit of the Royal Netherlands Navy, Lips Propeller Works employed him, initially as an industrial scientist and later in various managerial positions. In the shipbuilding group IHC Holland, he was managing director of their shipyard Gusto, specialised in offshore equipment. He served as director of corporate planning and R&D in the Royal Boskalis Westminster Group. In the late 1980s, he founded the innovation company Van Gunsteren & Gelling Marine Propulsion Development for the further development of his invention of the slotted nozzle (duct with a slot at the front), which ultimately led to the successful application of the wing nozzle (duct with a slot at the rear). He served on various boards for monitoring R&D subsidies, among others as vice chairman of the board of the Dutch Foundation for Technical Sciences 'STW'. Since 1997, he lectures, at Delft University, computer-aided support in architecture, urban planning, and project management. His publications include eight patents and ten books. His latest article was published in 2020.

 **Arnold G. Vlas** (1959) is an international business executive and entrepreneur with over three decades of experience in product management, marketing, and business development in various assignments in Europe and Asia Pacific. He graduated in the Faculty of Business Administration of the Erasmus University Rotterdam on a project in Taiwan where he stayed in the first few years after his graduation. His first job in the computer industry was in Taiwan's Acer Incorporated, as country sales and distribution manager for Korea and Hong Kong. A few years later, he joined Intel Semiconductor Asia Pacific in Hong Kong to become Director of Corporate Marketing and member of the regional executive staff in Asia Pacific and later Europe, Middle East, and Africa. Arnold and his teams were key contributors to establishing the Intel® and Pentium® brands as household names and IT manager check-off items, and successfully unlocked Asia Pacific as Intel's highest growth market, and maintained Europe as its most profitable region. During his last few years at Intel, he joined and co-led a global Mobile Product Development team based in Shanghai. More recently, he transitioned into the world of academia, built on his extensive training and organisational development work at Intel and as a consultant, and started teaching at universities on topics like product management, innovation, marketing, and sustainable business.

Abbreviations

c/D	Chord/diameter ratio
CCP	Chinese Communist Party
CEO	Chief Executive Officer
CFO	Chief Financial Officer
COO	Chief Operating Officer
CPP	Controllable pitch propeller
CRP	Contra rotating propellers
HBR	Harvard Business Review
HR	Human Resouces
IP	Intellectual Property
IT	Information technology
LG	License Giver
LT	License Taker
MIC	Made In China
NIHS	Not invented here syndrome
OEM	Original equipment manufacturer
PC	Personal computer
P&L	Profit and Loss
RD&D	Research development and demonstration
RDM	Rotterdamse Droogdok Maatschappij
RSV	Rijn-Schelde-Verolme
VDI	Virtual desktop infrastructure
WTO	World Trade Organization
WWII	World War Two

Introduction

1

Innovation is a polluted term. When an audience like a class of MBA students or a group of managers is asked to describe what the term means, one gets widely diverging answers. Literature on the subject also shows a variety of definitions. Most of them point at *novelty* and *application*. For instance, Mansfield (1968, p. 83) writes: *'An invention, when applied for the first time, is called an innovation'*. This definition covers most of the literature on technological innovation and is, therefore, adopted throughout this book.

Similar ambiguity can be noticed in regard to the related terms *invention*, *quality*, *uncertainty*, *risk* and *diffusion of innovation*, as will be explained later. As a result, practitioners get confused and benefit only piecemeal from the vast amount of literature on these subjects.

Practitioners involved in technological innovation seem to live in a world separate from the world of scientists who write about it, often without ever having been involved in innovating endeavours themselves.

As Peter Drucker states, the goal of industrial R&D is to innovate, not merely to invent (Drucker 1985, p. 127). This has been the starting point of van Gunsteren's book on industrial R&D (van Gunsteren 2003a). The third revised edition being out of stock made him decide to write a more comprehensible volume targeted at a wider audience, in particular politicians, civil servants and captains of industry and their staff engaged in innovation-related strategic decision making.

The Covid19 crisis outbreak forced both authors to stay at home for quite some time, which offered the opportunity to make this a joint effort, based on the background of van Gunsteren in the realm of maritime innovations and the experience of Vlas in the computer industry. The cases and examples are predominantly from these fields, since we have limited the scope of the book to insights that are corroborated by our own experiences and observations from these two fields.

Starting point of the book is the License Giver business concept described in the next Chapter and Appendix A.

L. A. van Gunsteren, A. G. Vlas, *The License Giver Business Concept of Technological Innovation*, Future of Business and Finance, https://doi.org/10.1007/978-3-030-91123-2_1

Our first assertion is as follows:

Once you have achieved design leadership, be prepared to license the design out in one form or another at an appropriate price. Never aim at a monopoly in which everybody excluded from it becomes your adversary instead of a potential partner. Licensing should always be taken into consideration as a strategic option.

Our second assertion is as follows:

The paradox of management of innovation is that innovation cannot be managed directly. Innovation requires creativity and inductive thinking that cannot be imposed top-down (Appendix C). It is useless to order someone to be creative. The motivation to be creative has to come from within. All one can do is making circumstances conducive to innovation by measures that facilitate, but by no means guarantee, innovation to happen.

Innovation, defined as the first application of an invention—the first evidence that someone is prepared to pay for it—is a *game of excellence* since only one can be the first. An inner drive to excel is a prerequisite for the successful conversion of inventions into innovations. Like in top sport, excellence—being the best—requires *intrinsic motivation to contribute to a common goal*, be it winning a gold medal in the Olympics or the creation of products that are really contributing to technological progress and, ultimately, prosperity and wealth.

Staying at the top represents an even greater challenge than getting there.

Business success contains the seeds of its own destruction (Grove 1997, p. 3). The more successful you are, the more people want a chunk of your business and then another chunk and then another until there is nothing left. Other people's attacks are inevitable. In technology, whatever can be done will be done. To remain successful, the corporation has to engage in new innovative endeavours.

Commercialisation of the current generation of an innovation and the parallel development and launching of next generations is a fragile process.

Such parallel development requires maintenance of the innovative culture that enabled the first generation to be created. This brings along the need of a disciplined execution of a commercial strategy that strives to achieve two equally important License Giver objectives: financial returns as well as global design leadership across multiple product generations. Achieving these objectives requires behaviours that feel unnatural, particularly licensing out the most advanced technology to competitors instead of protecting and keeping technology and production in-house. The timely obsoleting of one's own successful products instead of extending its market presence and 'milking' it is essential for continued success.

The game of technological innovation, the License Giver business concept, can be of great help in the structuring of the efforts required for any realistic innovation ambition.

The License Giver Business Concept

The two conditions for successful technological innovation, excellence and intrinsic commitment to a common goal constitute the basis for the classification of business identities in the strategic business concept of *License Giver*, *License Taker*, *Jobber*, *Consultant*, briefly referred to as the *License Giver business concept*.

The article 'Planning for Technology as a Corporate Resource: A Strategic Classification', Long Range Planning, 1987, describes the License Giver business concept in detail (reproduced in Appendix A). The concept is extended here by including *design quality* as a guiding principle in the License Giver's strategic decision making.

What characterises a good engineering design?

Some exemplary designs spring to mind that have earned a reputation of being excellent designs for the era in which they were conceived.

Ships
- Viking ships—The Viking ships enabled the Vikings to discover America a long time before Columbus.
- VOC ships—The VOC ships could be built so quickly, that the VOC could grow to be the largest corporation in the world.
- Liberty ships—They could be built by laymen, ultimately at a rate of one per week. The German submarines could never sink the ships at such a fast rate.

Aircraft
- Boeing 747 Jumbo Jet—The Boeing 747 had exactly the right increase in size at the time required by the market, although there were no new technologies involved in the design.
- Spitfire—When Göring asked his air force general what he needed to win the war in the air, his answer was 'Give me a squadron Spitfires'.

© The Author(s), under exclusive license to Springer Nature Switzerland AG 2022
L. A. van Gunsteren, A. G. Vlas, *The License Giver Business Concept of Technological Innovation*, Future of Business and Finance,
https://doi.org/10.1007/978-3-030-91123-2_2

Vehicles
- 2 CV—Low-cost car in which farmers could transport eggs without breaking them.
- Porsche 911—At the time, a breakthrough in sports car performance with the engine placed at the rear.

Military equipment
- All welded Russian WWII tank—This tank, developed by Paton in the Ukraine, terminated the German superiority in WWII tank battles.

Buildings
- Eiffel Tower—An eye catcher for the world exhibition in Paris which is even today a symbol for the city.
- Golden Gate bridge—The symbol for San Francisco's progressive modern society.
- Sydney Opera House—Ideal acoustics in a building which became a symbol for the whole country.

What do these and other brilliant designs have in common?

First, *fitness for purpose*. The design satisfies the requirements that follow from its mission exceptionally well.

Second, *technological balance*. All subsystems and components are in balance with each other: levels of reliability, sophistication, luxury, cost, etc. are all in the same range. In other words, a characteristic of a good design is that it does not have much wasted quality (van Gunsteren 2013). Wasted quality is nihil in the engineer's ideal of Caesar's war chariot, which never fails but at the end of its lifetime disintegrates completely into dust. If one bolt were to remain, then that bolt would have been constructed too conservatively and that would have had adverse weight implications. Unnecessary weight impairs the effectiveness of the chariot, which Caesar would never have accepted.

Third, *state-of-the-art technology*. Available technologies that can effectively be used are indeed applied. Appropriate, opportunistic use is made of state-of-the-art technology that has proven itself.

These three characteristics of a good engineering design make designing largely a skill, requiring a way of thinking as well as knowledge of relevant technologies and methods (van Gunsteren 2000).

2.1   Classification of Business Identities

To establish the identity of an organisation as far as exploiting technology is concerned, two questions are of particular importance:

1. Can the organisation be characterised as one of *doers* or as one of *thinkers*?
2. Is the organisation offering a *product* or a *capacity* to make something on specification?

The four possible combinations of answers to these questions can be placed in a matrix (Fig. 2.1).

The four quadrants have been labelled:

- License Giver
- License Taker
- Jobber (high or low technology)
- Consultant

License has to be taken here in the broadest sense of the word, comprising much more than a mere legal Licensor-Licensee agreement. A License Giver may not give licenses or even take them on certain components or subsystems, while nevertheless remaining responsible for the worldwide design leadership of the product.

Successful firms tend to fit in just one of the four quadrants or have separated their organisational subunits in such a way that each one clearly fits into only one quadrant. Strategic dilemmas and organisational stress tend to occur when the different characteristics associated with each of the four business identities simultaneously appear within one organisational unit. This can be explained by comparing the features of the four identities (see Table A.9 of the Appendix A).

The essence of the License Giver business concept is to keep the four business identities separated in fairly independent organisations that can develop their own norms and values. When executed well, true to their business identity, these four identities can be the basis of equally successful businesses from a financial perspective.

Mixing them in one organisation, being stuck-in-the-middle, leads to identity crisis (Fig. A.7 of Appendix A). For instance, the License Giver's guiding principle is *effectiveness*, while for the License Taker it is *efficiency*, and for the Jobber *occupancy* of a general-purpose facility.

Fig. 2.1 Classification of business identities

	Doing / Making	Thinking / Knowking
Product	License Taker	License Giver
Capacity	Jobber (High or Low Technology)	Consultant

Case: Dell, from Disruptive License Taker and PC Market Leader to a Stuck-in-the-Middle Identity Crisis

After Michael Dell started assembling pc's-to-order from his dorm room in 1983/1984 on a small scale, he burst onto the bigger PC market with his disruptive 'Direct-Sale and Build-to-Order' supply chain model in 1986. Timing was perfect for servicing the maturing US 'replacement and upgrade' business market segment. In its first year of operation the company already achieved revenues of US$ 70 Million. The consumer market soon followed with the users buying their second or third PC, being more price-conscious, more PC experience, and comfortable buying over the phone. Dell continued to tune their efficient lean direct model, adding e-commerce in the mid-nineties, while competitors HP, Compaq, AST, Packard Bell could not follow without jeopardising sales by alienating their distributor and retail channels. By the turn of the century Dell's operating cost, as a percentage of sales, ran at less than half of major competitors. Michael Dell became the youngest CEO of a Fortune 500 company on record in 1992 and with surpassing Compaq Dell reached the first place in global PC sales in 1999.

To continue to grow, Dell expanded its product line with servers, peripherals, etc., but competition started to close in. The continued commoditisation of PC's and related margin pressure drove Dell to look for new and better business opportunities. IBM had successfully transitioned into a software and services business by acquiring PWC's advisory and system integration business, and in 2008 HP followed by acquiring EDS for US$ 13.9 Billion. Dell tried to follow by acquiring Perot Systems for US$ 3.9 Billion in 2009. The press announcement stated that the intention was to leave the consultancy operation largely independent under its existing management team. Instead of living up to that intention, Dell's management soon started to get more actively involved. The result was a major clash of two different business identity cultures (Fig. 2.2).

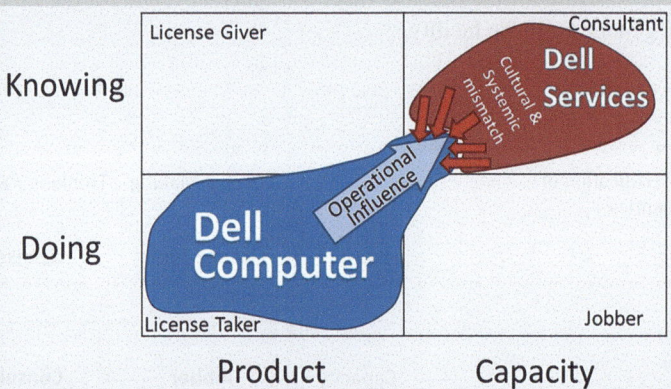

Fig. 2.2 Dell's Culture clash: managing a Consultant capacity firm with the mindset of a Product License Taker firm

Dell's License Taker culture was a terrible match with that of the newly acquired 'Dell Services' consultancy business.

An exodus of senior staff followed. Dell then realised that the work cultures were incompatible. The mix had made 'the whole less than its parts'. Dell cut its losses by selling the Dell Services unit to NTT at a loss of nearly US$ 1 Billion, only a few years after the acquisition.

Apparently, the difference between a product firm and a capacity firm is not well-understood (Simon, 1989). The following hypothetical example serves to explain the principle:

Suppose you want to buy a pair of orange leather shoes with a rosette on it. You search the internet but cannot find what you are looking for. Leather shoes are offered in colours brown or black and with or without laces. In order to get what you want, you go to a leather shop and ask the owner if he can deliver a pair of orange shoes with a rosette and how much he would charge you for it. The shop owner assures that he can deliver, possibly with a choice of leather qualities and likely at a higher price than any of the shoes from your internet search. Since you are willing to pay extra for getting exactly what you want, the deal is closed.

What did you buy? Not a product, but a part of the shop owner's capacity to make a variety of leather goods, be it a pair of shoes, a pair of boots, a leather chair or a leather belt.

Managing a capacity firm with the mindset of a product firm is heading for disaster.

Case: History of the IHC Gusto Shipyard
During about a century, the founder's family Smulders had successfully managed the yard. The mission as can be inferred from the history of projects that were executed was to offer the *capacity to make anything out of steel, including designing and assembling*. The yard's track record shows a wide variety of building numbers: cutter dredgers, trailers, cranes and crane ships, bridges, docks, ferries, jack-ups, semi-submersibles, drill ships, single buoy mooring installations, tugs, coasters and many more. In the last few years before the yard's closure in the late seventies, the Executive Board of IHC displayed a mindset of a product firm which clashed severely with the historical capacity identity of the yard. The result has been a whole series of erroneous decisions, such as cancelling the investments for increasing lifting capacity and working under a roof. The Executive Board publicly announced never to engage anymore in building semi-submersibles. By voluntarily limiting their scope to a market segment of low demand, they actually defined themselves into the grave.

2.2   Design Quality as the Guiding Principle for the License Giver

As mentioned before, the License Giver–License Taker relationship entails more than a mere legal Licensor–Licensee relationship. The License Giver must maintain control over the quality assurance of the License Taker. *Quality assurance* is the set of rules and regulations to define and ensure quality of a product (ownership lies with the License Giver). *Quality control* relates to the fulfilment of the rules and regulations in operations (ownership lies with the License Taker). The License Giver ensures that the product's quality, fitness for purpose, is maintained worldwide.

Quality, thus defined, is the main criterion for decision-making in the License Giver organisation, not shareholders' value or stakeholders' interests that are subject to change in due course. Quality, once vested in the product, is of a more lasting nature.

The goal is to get execution as much as possible in line with what is needed for fitness for purpose, not necessarily what is prescribed in rules, regulations and specifications. When necessary, there is always the possibility of applying for an exemption from a rule or specification.

Quality control and assurance according to the mainstream of literature is getting execution in line with design *specifications* (Fig. 2.3).

In the License Giver business concept, *fitness for purpose* is adopted as the guiding principle (Fig. 2.4).

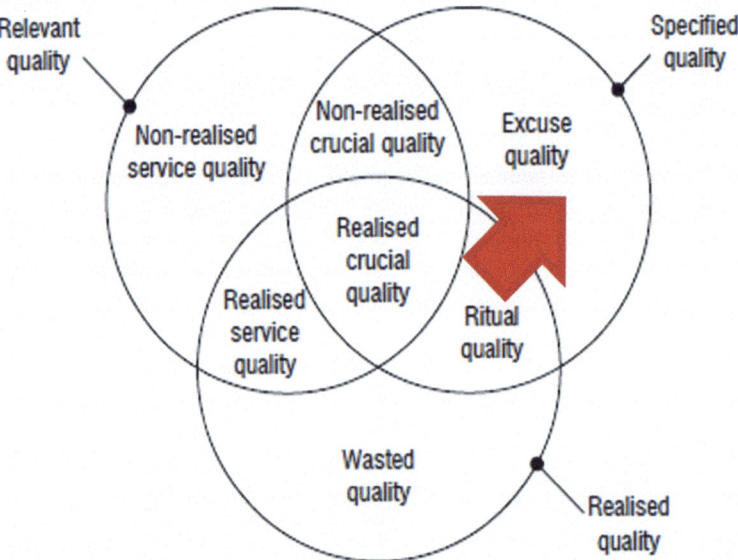

Fig. 2.3 Emphasis of quality control and assurance according to the main stream of literature: getting execution in line with design specifications

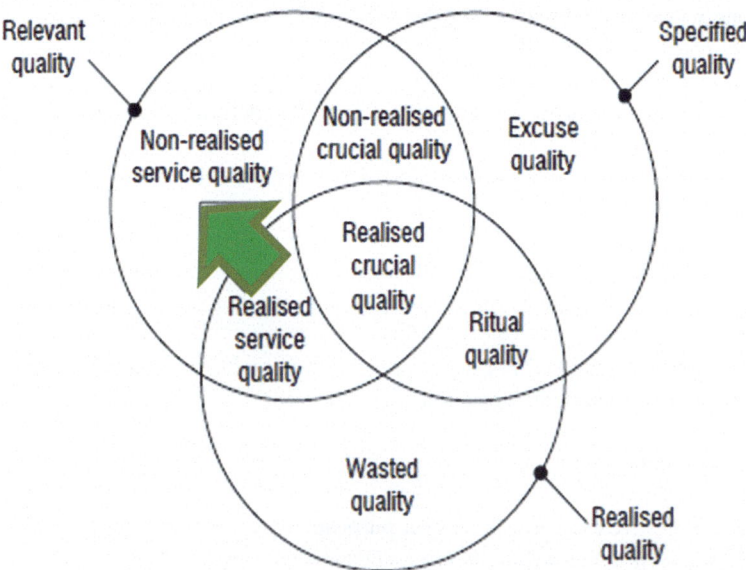

Fig. 2.4 *Fitness for purpose* guiding criterion in License Giver business concept

These assertions are based on our classification of seven categories of quality described below (van Gunsteren 2013).

2.2.1 Classification of Seven Categories of Quality

What is quality?

Doing or making something well according to the norms of an evaluator or user.

These norms depend on the purpose one has in mind, hence the definition:

Quality is fitness for purpose.

That means quality is

1. Related to a subjective purpose.
2. A perception.

Absolute standards of quality do not exist. What quality depends on the needs of the user. These needs are not only determined by the user's personal desires and preferences, but whenever new technologies offer new possibilities, the wishes of users will also become more demanding.

If one wishes to get something done from a larger group of people, one has to resort to regulations: laws for a country; rules and standards for a trade; rules, procedures and policies for a corporation. Therefore, quality is not only a matter of knowledge and mentality, but equally of a proper definition of adequate quality specifications.

Quality specifications, norms enabling the measurement of performance in doing or making, depend on:

1. Purpose of the user (e.g. clean office, car that does not break down, etc.).
2. Experience in the past as far as user problems are concerned (breakdown, wear and tear, etc.).
3. What can be measured? For instance, environmental rules should not be so strict that violation cannot be measured.

Quality can be:

1. Relevant or irrelevant for fitness for purpose.
2. Realised or not realised in the product or service.
3. Specified or not included in specifications.

Combinations of these aspects yield seven categories of quality which we will now discuss.

Quality specifications will never cover exactly all quality which is relevant to the user (Fig. 2.5).

Relevant quality which is covered by specifications is labelled *crucial quality*, because it is absolutely crucial to realise this type of quality in the product or service. In the case of non-compliance, a claim would be justified both formally and because the user really needs that quality for his purpose. Relevant quality which is not specified is called *service quality*, because this quality has to be delivered as a service if the user's needs are to be properly satisfied. Specified quality that does not serve any purpose of the end user is labelled *cosmetic quality*.

Cosmetic quality consists of:

1. *Ritual quality*: realised cosmetic quality
2. *Excuse quality*: non-realised cosmetic quality

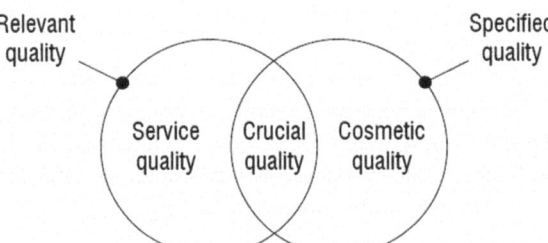

Fig. 2.5 Quality specifications never cover exactly all relevant quality

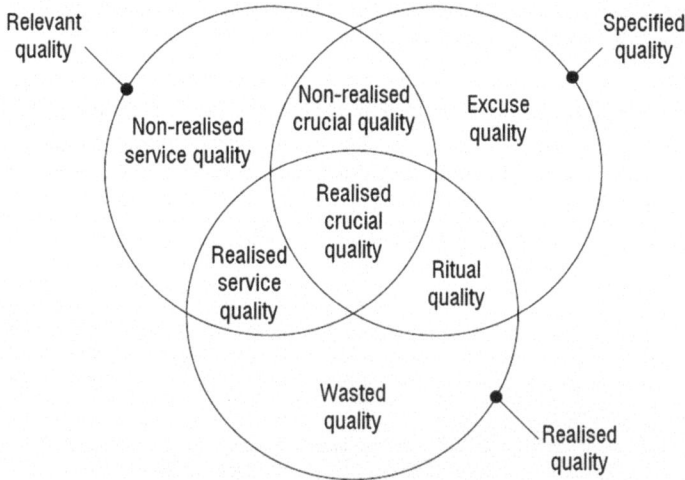

Fig. 2.6   Classification of seven categories of quality

Specifications (related to rules and regulations) are sometimes used as an excuse to exclude a supplier. For instance, the dimensions of car number plates in a certain country were prescribed in such a way that foreign suppliers were handicapped. In another country, an old-fashioned, inaccurate method to measure the dimensions of marine propellers (using templates) was prescribed to protect the backward domestic industry against more advanced international competitors.

Cosmetic quality should not be confused with cosmetic measures to give the product an attractive appearance, such as good-looking packaging. This kind of cosmetics belongs to service quality, as it satisfies a real user's need.

Realised quality which is neither relevant nor specified is labelled *wasted quality*, as it serves no true purpose.

This completes our classification of the seven categories of quality (Fig. 2.6).

The relevance of paying attention to fitness for purpose is illustrated by the following cases.

Case: Flyover
A flyover construction project was seen as a success by the Ministry of Infrastructure, since it was completed according to the contractual specifications, on time and within budget. The purpose of the project was to remove the daily traffic jams during rush hours. That purpose, however, was never achieved. The traffic jams only shifted a few miles away from the original location. The 'success' project was actually a waste of money.

Case: Brienenoord Bridge Rotterdam

A similar case concerns the Brienenoord bridge connecting the North part of Rotterdam with its Southern area. The bridge was commissioned in 1965. Its capacity soon turned out to be far insufficient to avoid daily traffic jams. In the end it was decided to add an identical bridge adjacent to the existing one. The commissioning of the second bridge took place in1995. Figure 2.7 shows a photograph of the two adjacent identical bridges.

Fig. 2.7 Picture of the extension of the Brienenoord bridge Rotterdam. (Source: Rijkswaterstaat (RWS) beeldbank)

Case: Entrance Poster Shell Office Houston, Texas, USA

At the entrance of the Shell office in Houston, Texas USA, its visitors are welcomed with a company poster with the text 'Definition of Quality: Conformance to Requirements' (Fig. 2.8).

Fig. 2.8 Poster in Shell office in Houston Texas

Quality is definitely not equivalent to conformance to requirements, as illustrated by the three cases above. Apparently, the myth of compliance is deeply rooted in the construction industry.

Our assertion is that fitness for purpose should be the guiding principle for the Jobber. The use of the seven categories of quality classification in the construction industry is described in detail in Appendix B.

2.3 Innovation and Renewal

Since people hold widely diverging ideas on what innovation really means and in what sense it distinguishes itself from related notions such as invention and discovery, we need to provide some definitions (Table 2.1, van Gunsteren 1987).

Innovation relates to the first use—the first commercial application—of a new product, concept or idea. The recognition of its usefulness by a user generates a change in the socioeconomic environment. It is this utility function that distinguishes innovation from discovery and invention. Discoveries and inventions do not have socioeconomic value unless they serve as cornerstones for an innovation. Diffusion of innovation relates to the rate of adoption in the marketplace.

The innovator applies something new for the first time. It is new not only to himself but also to everyone else in the world. Someone adopting an existing innovation that is new to him but not to others is not an innovator. He is an imitator, who applies what is already available outside his organisation. For that organisation itself, however, the effects are in many respects similar. Innovations generated by the

Table 2.1 Some definitions related to innovation

Discovery	New Insight: Finding something useful. Example: Newton's Law.
Invention	New technical trick for which, in principle, a patent can be granted but which might be utterly useless. Example: Automatic hat lifter, a patented mechanism allowing to greet somebody by lifting the hat without having to touch it.
Innovation	First application of something new: first evidence that some user recognises its usefulness and is prepared to pay for it.
Diffusion	Adoption of the innovation in the market.
Renewal	Adoption of innovation seen from within, introducing an innovation that is new to the organisation but not to the outside world.
Improvement	Marginal innovation, an operational improvement that, although original and useful, is of such marginal nature that it hardly generates strategic impact.

organisation itself as well as adoption of existing innovations disrupt the status quo and affect the balance of power and influence in the organisation. Innovation and adoption of existing innovations tend to be confused by managers who perceive them as a single issue: incorporating available technical know-how into their organisation.

To avoid this confusion, we introduced the term *renewal* to refer to *all* available technical know-how to the organisation regardless its origin. When we speak of diffusion of innovation, we focus on the innovation. Renewal, in contrast, emphasises the newness to the organisation, which is what matters to its management.

The distinction between innovation and renewal is fundamental. Innovation is a game of excellence, within reach of only a happy few. They are the first, others will have to follow. To see innovation as a necessity for all firms is as absurd as pleading that all sportsmen should participate in the Olympics. Innovation is not necessary for all firms, but renewal is. When competitors achieve cost savings by using word processors instead of conventional typewriters, one has no other option than to follow. Only if the firm pursues a strategy of design leadership in text processing devices, there is a need to innovate. The need to innovate, therefore, depends on the firm's identity in regard of its exploitation of technology (Fig. 2.9).

The *creator* of new technology *differentiates* himself through innovation, using the life cycle concept and effectiveness—not efficiency— as governing principles.

The *user* of new technology *imitates* through renewal, using the experience curve and efficiency as governing principles.

In short, technological innovation is just one way to make money. Innovation is only essential for License Giver type organisations. Their main goal is excellence— to be the best—which is only within reach of a few. Renewal is necessary for all firms. The name of the game is to exploit what is already available. Consultants can be of great help in this matter (Table 2.2).

The pace at which diffusion of innovation proceeds depends on the technology and related commercialisation strategy in the particular business environment involved.

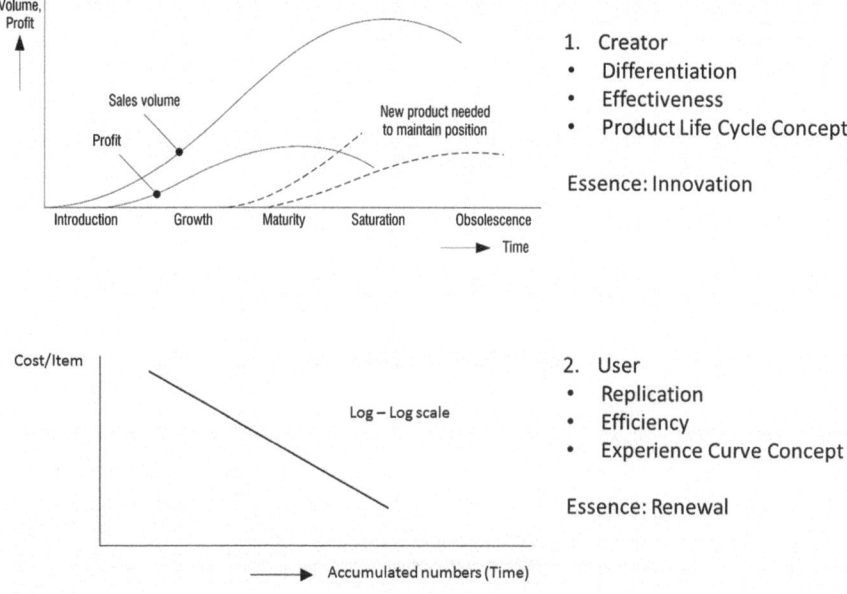

Fig. 2.9 Identity related to the use of technology

Table 2.2 Emphasis in innovation and renewal

Innovation	Renewal
• Depth, only excellent is good enough	• Breadth, good is good enough
• Self, unique knowledge and working method	• With consultant's assistance, knowledge is for sale

Many countries have established government or semi-government technical institutes—like Fraunhofer Gesellschaft in Germany, TNO in the Netherlands, and ITRI in Taiwan—that are supposed to support innovation in the country. The actual mission of these institutes, however, is not to innovate, but to provide engineering consultancy services to the industries of the country, that means to assist companies in their implementation of renewal.

2.4   The Importance of Location

The License Giver organisation needs to be located in a cluster of its particular industry. Clusters are critical masses—in one place—of unusual competitive success in particular fields. Famous clusters are the computer industry cluster Silicon Valley, USA and the automotive industry clusters in München and Stuttgart, Germany (Porter 1998). Enduring competitive advantages in a global economy lie in local things—knowledge, relationships, motivation—that distant rivals cannot match. Clusters affect competition by:

Case: Client X, Developer of a Virtual Desktop Infrastructure (VDI)
Solution, Taiwan

The performance and features of the new VDI management solution, as demonstrated in the lab, had true innovative potential. The VDI solution represented a drastic departure from the then-dominant client-server architecture designed around devices with powerful processors and storage capacity. The basic idea of the VDI solution was to limit the PC to input and output functions, wireless connected via internet to the server which conducted all computations (Fig. 2.10).

VDI solution concept

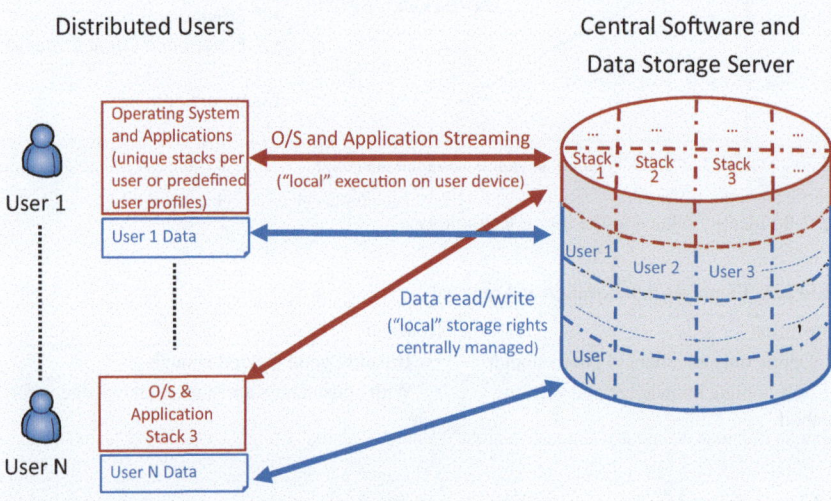

Fig. 2.10 Basic idea of the new VDI solution

The authors were engaged to help develop the business strategy and action plan to commercialise the invention.

A prototype had already been developed and patents were pending.

The authors were engaged to assist in developing a business strategy and action plan to commercialise the invention.

The recommended strategy was to aggressively pursue *direct sale* into the education market. Schools and universities were looking for VDI systems to support their administration as well as student computer labs. Its existing IT infrastructure was less complex compared to other market segments. Schools and universities were also expected to be a more forgiving buyer group. They formed a good early adopter market segment to test the solution in real-world environments, and to work out the inevitable teething problems of new technologies. In this way the education market would serve as a good source of reference customers and demonstration sites for commercial partners and other market segments.

For the other market segments, especially government and large private enterprises, the authors recommended an aggressive *licensing strategy*. These large companies were highly sensitive clients with a legacy of complex and fragmented IT systems, very different from the environments in the company's labs. Adopting and transitioning to an organisation-wide VDI architecture represented not only a significant investment, but also an operational risk. In addition, these companies had long existing relationships with the established 'system integrator' companies HP, IBM, Perot Systems, Capgemini, which had been selling and installing their existing systems. Entering into a licensing relationship with a few of these large integrators would provide a faster and safer entry into this main market. Also, the authors thought it to be the only approach that stood any chance to reach the scale required for becoming a leading global design standard. The licensing strategy would also pave the way to potentially sell the solution or company to one of these licensees as an investor end-game strategy.

The President and Founder, however, did not accept the recommendations. He held an overly optimistic view about his solution's technical readiness to go prime time inside these unforgiving segments, as well as the ease with which his VDI solution would be broadly recognised as the 'revolutionary and obvious way to go' to replace current IT installations. Furthermore, a key personal objective of the President seemed to build a large company himself (possibly to prove himself towards the family he had married into, which presided over a large conglomerate of successful companies), as opposed to an open-minded view on finding a path with higher probability of establishing the technology as a global standard and optimal investor end-game scenarios. Under-appreciation of the efforts, creativity and discipline required to commercialise an invention, and the unwarranted expectation that the invention 'will sell itself' is a common pitfall for start-ups.

Six years after the authors finished the engagement the company had still not generated any significant revenue, nor established any significant 'License Taker' partnership, and soon thereafter joined the annals of promising inventions that remained unrealised innovations as a result of poor commercialisation.

1. Increasing the productivity of companies in the area.
2. Driving the direction and pace of innovation.
3. Stimulating the formation of new businesses, which expands and strengthens the cluster itself. A cluster allows each member to benefit as if it had greater scale or as if it had joined with others formally, without requiring it to sacrifice its flexibility.

A nearby university is of particular interest since it can offer interesting opportunities for joint R&D-projects, apprenticeships and, above all, allowing high-flyers to be spotted long before their graduation.

Case: The Silicon Valley Computer Industry Cluster
In the late-fifties and sixties, Northern California became the cluster for a new breed of technical and business leaders: Bill Hewlett, Dave Packard, Bob Noyce, Steve Jobs and many others. The universities of Berkley, Stanford and Caltech developed ancillary facilities such as the Stanford Industrial Park, set up to stimulate collaborative efforts with private industry. After the investment by Arthur Rock in Fairchild Semiconductor, a vibrant hi-tech Venture Capital community emerged together with a new employee-ownership compensation approach, adding another key ingredient that formed the incredibly innovative 'Silicon Valley' industry cluster.

2.5 The License Giver Business Concept Applied in Practice

The License Giver concept can be seen as an idealised design for an innovating organisation.

An idealised design, in German language referred to as Leitbild, describes a desirable future that may never be reached but toward which progress is believed to be possible (Ackoff 1981, pp. 104–125).

Case: Sulzer, Diesel Engines
The world's largest producer of diesel engines for ocean going vessels in the latter half of the last century was located in Winterthur, Switzerland, a country far away from any ocean (Huibregtsen 2017). Sulzer was an extreme type of License Giver. Only 5% of its revenues came from production of its own-design cylinders; 95% came from License fees.

When the in-house production becomes small in comparison to the licensed production, the profitability of it becomes of secondary importance. The in-house production is then only maintained as a means to receive direct operational feedback and serve as a test facility for improvements in the design. Its relatively high production cost constitutes a price that has to be paid for maintaining design leadership and the continuous demonstration of the License Giver's superiority in design technology.

Sulzer's transition towards its favourable position of an extreme type of License Giver was made possible by its original position in land-based diesel engines. The inventor of the diesel engine, Rudolf Diesel, had actually worked at Sulzer as a trainee long before inventing the engine. With the ETH Zürich on its doorstep, its location was in the centre of the Swiss cluster of precision engineering technologies.

Technical know-how can only to a limited extent be protected by patents. Practical experience, thoroughly tested computer design programmes, and in particular a track record of reliability, constitute valuable technical know-how that cannot be protected by patents. How can such know-how be transferred to other parties with an appropriate financial reward for the supplier of that know-how? A *technical assistance contract*, specifying a percentage of the turnover of the License Taker to be paid to the License Giver, can be highly beneficial to both.

Case: Lips Propeller Works, Overseas Companies and Joint Ventures for Marine Propellers
Lips Propeller Works, a leading manufacturer (License Giver) of ship propellers located in The Netherlands, had in due course after WWII expanded its overseas business in a number of countries by means of fully owned as well as 50–50% Joint Ventures (License Takers). For instance, in Italy, propellers were produced and sold by the JV (License Taker) in Italy. The partner in the JV was a consortium of Italian shipyards that controlled 90% of the Italian market. The Joint Venture agreement included a clause that 10% of the turnover had to be paid by the JV to the License Giver in The Netherlands for technical assistance, who would provide all necessary technical support and, above all, guarantee quality to customers. The latter was essential, because without such guarantee many ship owners would refuse to buy propellers produced in the Italian plant. If the Italian managing director would give unnecessary discounts to Italian shipyards, reasoning that of each dollar discount $0.90 would go to the Italian JV partner instead of $0.50, the License Giver JV partner would not be bothered about it. The License Giver had to deliver: the plant manager, the propeller designs, and expertise in quality-related issues. The cost thereof being less than 1%, left the License Giver with a handsome net license fee of 9%.

Case: Campina Dairy Company in Friesland, Cheese Products
Cheeses used to be developed and produced in four plants in the province. Development and production were then separated according to the License Giver concept: one License Giver where new cheeses were developed with emphasis on effectiveness and three License Takers for production only with emphasis on efficiency.

The Intrinsically Motivated Crowd

3

3.1 Lessons from History

Groups of people can sometimes rise to impressive performance when they are highly committed to a common cause.

In WWII, Winston Churchill promised Blood, Sweat and Tears to achieve the common goal of victory. Commitment to that goal induced millions of people to make, voluntarily, great personal sacrifices including risking their lives.

The clarity of purpose and resolve of the British was not entirely unique during WWII, but the events on "May 13th, 1940, provide a uniquely illustrative moment in history:

GERMANS 13 MILES PAST LIEGE IN FLANK DRIVE; CLAIM TO HAVE CROSSED HOLLAND TO NORTH SEA; GREAT BATTLE NEARS AS ALLIES POUR IN FORCES

New York Times, May 13, 1940

Appointed Prime minister the preceding Friday, after forming a War Cabinet of unity, Churchill on Monday May 13th in his first address in the House of Commons:

"I have nothing to offer but blood, toil, tears and sweat.

We have before us an ordeal of the most grievous kind. We have before us many, many long months of struggle and of suffering.

You ask, what is our policy? I can say: It is to wage war, by sea, land and air, with all our might and with all the strength that God can give us; to wage war against a monstrous tyranny, never surpassed in the dark, lamentable catalogue of human crime. That is our policy.

You ask, what is our aim? I can answer in one word: It is victory, victory at all costs, victory in spite of all terror, victory, however long and hard the road may be; for without victory, there is no survival."

© The Author(s), under exclusive license to Springer Nature Switzerland AG 2022
L. A. van Gunsteren, A. G. Vlas, *The License Giver Business Concept of Technological Innovation*, Future of Business and Finance,
https://doi.org/10.1007/978-3-030-91123-2_3

Another example is the organisation of the Eleven Cities skating competition and tour over 200 km along the 11 towns of the province Friesland in the north of the Netherlands. The event is only possible in extremely harsh winters. Since the year 1909 when the first Eleven Cities tour took place, only 15 times the ice was considered to be thick enough. Some intervals between consecutive Eleven Cities tours have been more than two decades. This feature makes it a unique happening which large groups of people in the Netherlands, especially Friesland, feverishly await to be a part of and contribute to a successful staging.

Successful, accident-free, staging of the tour requires a large volunteer organisation to prepare the 200 km track within 2 days from the 'go' decision for the approximate 16,000 participants on the day of the tour (many more than the 11,237 athletes that took part in the 2016 Summer Olympics).

Inhabitants of the province offer, for free, lodging during two nights to participants from outside Friesland. Others offer warm water and healthy food along the 200 km track during the tour. It so happened that when unexpectedly some 500 m mats were needed for 'klunen' (walking on skates over land around short distances where ice is considered too thin) a call over the radio was enough to get people of the neighbourhood to offer mats from their homes.

The common goal of making the event a success inspires everyone to contribute where possible. Around the Tours held in '85, '86, and '97, there are plenty anecdotes of people in distant parts of the world, with only two days notice between the Tour go-ahead and start, dropping whatever they were doing to catch flights to the Netherlands to participate. And when asked, international top skaters facing the dilemma of having to choose between participating in the Olympics or the Eleven Cities Tour, the majority prefers the latter and would cancel their participation in the Olympics.

Similarly, the common goal of innovating endeavours to create new products that significantly contribute to technological progress inspires everyone involved.

The Eleven Cities Tour is a truly unique event, and illustrative of the type of intrinsic motivation at the heart of a healthy and sustainable R&D team effort. Over a dozen books on the Eleven Towns Tour have been published, but all are in the Dutch language. For this reason, an adapted English translation of the chapter "Impressions of a tour skater" from the book "The Fifteenth Eleven Cities Tour 1997 Heroic Battle in Icy Wind" is included in Appendix D.

The mindset of an intrinsically motivated crowd is a characteristic feature of the innovative organisation and its work culture.

Case: 'Shockley's Traitorous Eight', a Clash of Cultures
In the mid-fifties the founders of Intel, Robert Noyce and Gordon Moore, joined the brilliant scientist William Shockley at Shockley Transistor Laboratories. Shockley's reputation and the rumour of his Nobel Prize nomination enabled him to gather a number of hand-picked bright scientists to realise his vision of building one of the most innovative and successful semiconductor companies. Instead, while engineers and inexperienced businesspeople like Bill Hewlett developed their new 'Silicon Valley' management style, Shockley took his company into an entirely different direction. His distrustful and abusive 'East Coast' management style resulted in seven colleagues approaching the charismatic Robert Noyce in order to leave the company. Somewhat to their surprise Noyce agreed, and they left to become part of what Shockley referred to as the 'Traitorous Eight' (Fig. 3.1).

Under Robert Noyce's leadership the planar semiconductor manufacturing process was started up within a year. With the help of the newly arrived Arthur Rock, Noyce convinced the East-Coast company Fairchild Camera and Instrument Company to finance them and established Fairchild Semiconductor. The planar process near-instantly obsoleted the old transistor industry and helped to establish Fairchild as a License Giver. Fairchild now derived income from both its semiconductor products and licensing agreements with many of the old industry players. Under Noyce's leadership and Moore's technical brilliance Fairchild went on to develop the Integrated Circuit and became a record-growth company.

Less than a decade later Fairchild started to falter. An especially important factor behind its faltering was the mismatch in management culture between the hierarchical old-style East-Coast culture from the parent company and the emerging Silicon Valley culture that caused the loss of many of its best and brightest employees. Many of the departed employees established their own ventures or joined start-ups. They became known as the 'Fairchildren'. After attempting to save Fairchild the highly regarded leader Bob Noyce could not endure the East-coast culture and left together with Gordon Moore to form Intel, taking the driven young Hungarian refugee Andrew Grove with them.

Fig. 3.1 The "Traitorous 8", Gordon Moore, C. Sheldon Roberts, Eugene Kleiner, Robert Noyce, Victor Grinich, Julius Blank, Jean Hoerni and Jay Last. (1960, under the "Fairchild Semiconductor logo). (Source: Wikipedia (WP:NFCC#4))

That characteristic of an organisational environment cannot be imposed. Competent leadership creates an especially stimulating work environment. From the early sixties another feature of the success and strength of the Silicon Valley cluster was the emergence of a unique less-hierarchical and management style that contrasted with the more traditional US-East-Coast style.

In such a work environment, the individual team members tend to display loyalty to their profession and its technology leaders, rather than to the organisation they belong to.

3.2   Becoming a License Giver

All roads lead to Rome. There are several ways to become a License Giver:

1. From scratch, like Steve Jobs did when he started in 1976 to work with Steve Wozniak in the garage of a friend (Isaacson 2011, p. 21).
2. From the position of one of the other three basic identities, in particular the License Taker.

The transition has to be carefully planned and managed. The period that two different designs are offered and served in the market should be limited as much as possible to avoid a stuck-in-the-middle situation.

Case: Lips Propeller Works, Controllable Pitch Propellers Transition from License Taker to License Giver Identity

In the late sixties, Lips was Licensee of three different Licensors: Hundested Denmark, and two Dutch companies, De Schelde and Werkspoor. The considerable losses were financed by the profits from its Monobloc propellers. All three designs had serious disadvantages. The Hundested design involved a rod through a hollow shaft and grease lubrication. The design of De Schelde was limited to four blades placed in tandem. The Werkspoor design involved a rather large number of parts. Instead of upgrading their design Stork-Werkspoor sold their entire Controllable Pitch Propeller business including chief design engineer Jaap Wind to Lips. As soon as Jaap Wind joined Lips he started the development of a completely new design only assisted by a small team. A launching customer was found, concerning a retrofit that included a 16,000 HP CPP-installation. Its technical trial trip did not give rise to any problems. The next step was the delivery of two 35,000 HP controllable pitch propeller installations for a twin-screw container ship. Inside the company, the ship's technical trial trip evoked an atmosphere of tension (like in the NASA space trip for putting a man on the moon). Everyone in the company realised that the results of the trial trip would be decisive for the reputation of the company's own design. Great was the relief when only a minor problem emerged that could easily be resolved. The successful transition to the position of License Giver had taken about 3 years.

3.3 The Story of the Slotted Nozzle

Van Gunsteren's third patented invention was the slotted nozzle (British Patent Application no 44019, September 5, 1969). The history of the slotted nozzle is described by him in this section as follows (van Gunsteren 2004).

The patent description covers various combinations of nozzles and propellers: fixed nozzles, rotating nozzles attached to the blade tips—called ring propellers— and the combination of a ring propeller with a stator in front of it. In this section, only the slotted nozzle is discussed, that found full-scale application on two 20,000 HP seagoing tugboats (Fig. 3.2).

The slotted nozzle patent is a strong patent in the sense that it specifies the principle of a slot on a nozzle, thereby covering all possible variations on the shape and the location of the slot. Likewise, the well-known patent on the sail surfboard was a strong patent, since it covered all possible variations of a pivotal mast on a surfboard.

Extensive systematic open water tests were conducted for several slotted nozzle configurations at the Netherlands Ship Model Basin, MARIN, in Wageningen. The results were presented at the North American Tug conference (van Gunsteren 1973a) and included as Chap. 7 in van Gunsteren's Ph.D. thesis (van Gunsteren 1973b). The last conclusion of that chapter reads: The test results show that the

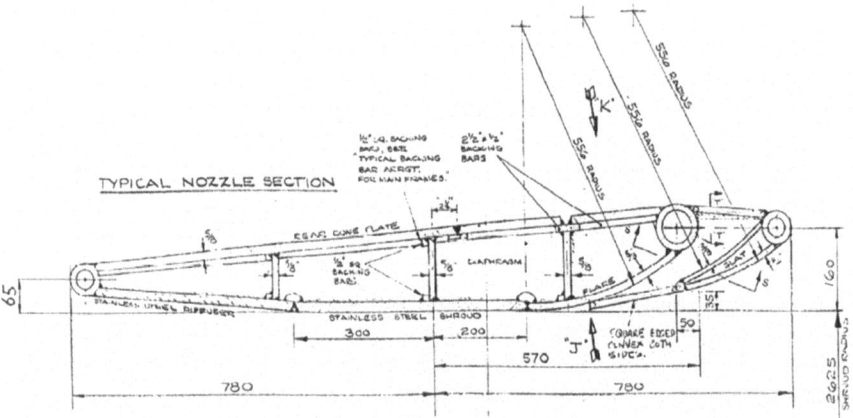

Fig. 3.2 Slotted Nozzle Profile for 20,000 HP Tugboats

slotted nozzle widens the field where shrouded propellers find their application, especially in those cases where efficiency at low thrust loadings (Thrust loading coefficient C_T of the order of 1.0) is important. The invention is of particular interest for double-duty ships like trawlers and tugs (see Fig. 3.3).

Not much later Lips received a request for proposal for the propulsion of two 20,000 HP tugboats for SAFMarine, Cape Town, South Africa, the 'Wolraad Woltemade' and her sister ship 'John Ross'. In free running condition at a speed of at least 20 knots, these tugboats would be sailing precisely in the region of $C_T = 1.0$ mentioned before. A bollard pull of 180 tons was required, which could only be achieved by fitting a nozzle. In view of the draught of the vessel, the diameter of the propeller within the nozzle could be 5.2 m at most.

It soon transpired that only the slotted nozzle could meet both requirements: a free running speed of at least 20 knots and a bollard pull above 180 tons. An open wheel would meet the free running speed requirement but fail to come even near to the required bollard pull. A conventional nozzle could produce a bollard pull over 180 tons, but in no way could reach the required free running speed of 20 knots.

Of the ten most powerful tugboats in the world at that time, Lips had only one, the 'Rode Zee', on its track record. The other nine boats all had controllable pitch propellers from KaMeWa, Sweden. With their 20,000 HP, the South African tugboats would be the most powerful tugs in the world.

The choice the client had to make between the two potential suppliers for the propulsion unit boiled down to:

1. KaMeWa: excellent track record, but a conventional nozzle with the consequence of a free running speed of hardly 19 knots.
2. Lips: limited, but good track record. The experience with the 'Rode Zee' was known to be flawless. The slotted nozzle enabling both requirements to be met, but also entailing the risk of a new concept. The slotted nozzle had only been tested at full-scale on a tugboat on the river Thames.

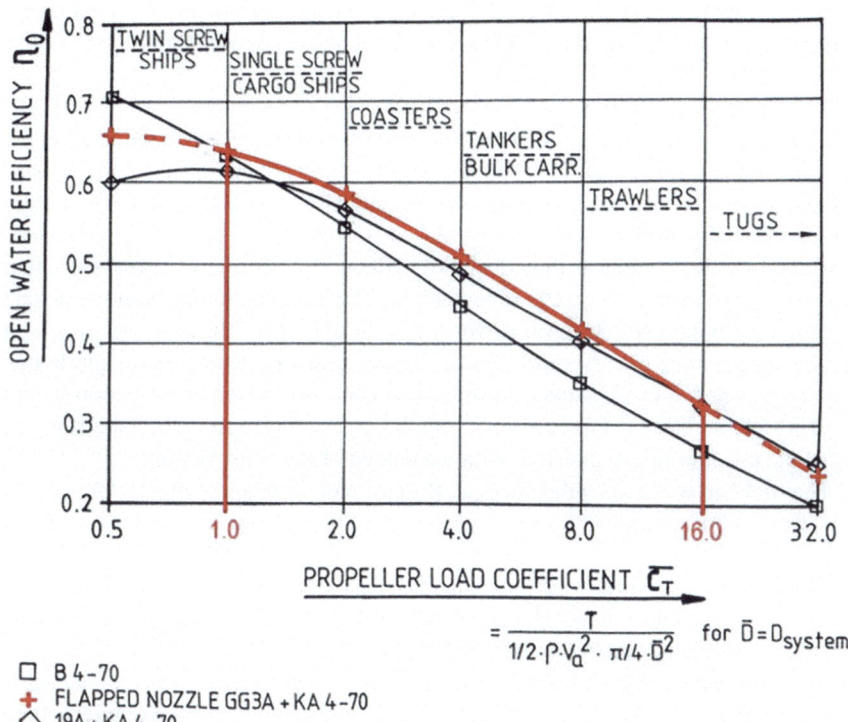

Fig. 3.3 Open water efficiency of the flapped nozzle (and slotted nozzle) with $c/D = 0.30$ compared with open wheel and conventional nozzle 19A

To resolve this dilemma, a meeting was arranged with the following participants: an engineer of the ship owner (SAFMarine), an engineer of the yard, the director of Kort Propulsion Ltd. (intended manufacturer of the nozzle), a propulsion consultant, a sales engineer of Lips and the inventor. They could not reach agreement. Whenever a conclusion came into sight, the consultant tabled a question that could not be answered straight away. For instance, how would a small change in the diameter affect the open water efficiency? A second meeting was scheduled a few weeks later. To prepare for that meeting, the inventor wrote a FORTRAN-computer programme in which the open water efficiency of all possible combinations of diameter, pitch, rotational speed (rpm), and speed of advance was computed in small steps. This was quite simple since the results of the systematic tests in the towing tank had been subject of a least squares curve-fitting procedure according to a three terms polynomial representation of three times (torque, propeller thrust, nozzle thrust) $3 \times 3 = 27$ coefficients (see pages 132 and 138 of van Gunsteren's PhD thesis, 1973b). When the propulsion consultant tabled his first 'what if'-question, the inventor put a 3-in.-high pile of computer print-out on the table and answered the question by turning pages until the requested combination was found. Subsequent questions were answered in the same way within minutes. When no more questions could reasonably be raised, the consultant resigned and agreed with the conclusion of the meeting that compliance with specific demands of the user should get priority in resolving the dilemma.

The client, accepting the risk involved, awarded the contract to Lips, and also accepted the recommendation that extensive model tests should be conducted in the vacuum tank at MARIN in Ede. The test results showed that the speed and bollard pull requirements could be expected to be met and also that cavitation would remain within acceptable limits.

The full-scale nozzle was manufactured by Kort Propulsion Ltd., London, at that time the market leader in nozzle manufacture. The construction of both the nozzle itself and its support structure, consisting of the head box and the heel, was specified in the contract to be the responsibility of our subcontractor Kort Propulsion. In this way, Lips' sales director wanted to limit downside risk. If the construction would fail, Lips would have a claim on Kort Propulsion. For this reason, the sales director urged the inventor not to interfere with the design of the construction.

The trial trip was a splendid success. Bollard pull with ample margin above 180 tons, free running speed well over 20 knots, excellent steering characteristics and no vibrations.

The performance in service was equally satisfactory, that is, during the first 6 months after commissioning. Then, one half of the nozzle broke off at the head box and the heel and disappeared to the sea bottom.

In the meantime, Max Lips, sole owner of the company, had fired the managers of the profit centres Monobloc propellers and Controllable Pitch propellers, both being sons-in-law heading towards a divorce. The inventor of the slotted nozzle was also fired because he was seen as part of a three-persons coalition against the two elder members of the Executive Board.

Jaap Wind, who was charged with resolving the problem of the broken nozzle, proposed to hire the inventor as a consultant, but the remaining two elder members of the Executive Board refused to do so. A short-sighted but understandable decision because the inventor had in the meantime started a lawsuit against the firm about the patent rights of the slotted nozzle. Without the involvement of the inventor, Jaap Wind had no other option than to play it safe and replace the slotted nozzle by a conventional one. As could be expected, the free running speed dropped by half a knot, a perfect, full-scale proof of the main advantage—free running speed—of the slotted nozzle over a conventional one. That the bollard pull slightly increased, must have been a poor consolation to the owner.

Many years later, Jaap Gelling made a thorough analysis of the breakdown. His vibration analysis showed that the nozzle's natural frequency had been in the region of the blade frequency of the propeller. The resulting resonance, which had caused the breakdown, could have simply been prevented by fitting struts to the hull at 4 o'clock and 8 o'clock positions of the nozzle circumference.

The lawsuit between the inventor and Lips about the patent rights of the slotted nozzle was based on Article 10 of Dutch Patent Law. The first part of that article states that if an inventor is employed in a function which brings along that he or she is expected to generate inventions, the intellectual property rights belong to the employer. The second part states, however, that the employer has to compensate the employee-inventor in accordance with the commercial value of the invention.

The lawsuit was by itself a lesson in patent exploitation. After 9 years and some Hfl 100,000 in legal costs the lawsuit was decided in favour of the inventor. Lips had to pay him Hfl 148,000 for the application on the two South African tugboats. Since Lips had argued that the invention was worthless because the nozzle could never be made strong enough, the court also decided that the patent rights had to be transferred to the inventor.

As this was the first time that an employee was rewarded on the basis of Article 10, II, the case (van Gunsteren versus Lips) has become jurisprudence in intellectual property legislation.

Jaap Wind's comments have been published as a contribution from an invited discusser (van Gunsteren 2004):

1. Historical data

   – 1973 Lips gets the order for two CPP-installations with (slotted) nozzles, each of 14.3 MW, for the most powerful tugs in the world. The customer is SAFMarine, Cape Town.
   – 1975 Delivery.
   – 1976 Commissioning. First 'Wolraad Woltemade', with sister ship 'John Ross' half a year later. On October 6th, after being in service for about six months, the 'Wolraad Woltemade' loses the port side of the nozzle. The 'John Ross' makes her trial trip. On my visit to SAFMarine, Cape Town, on November 11th–19th, I inspect both ships.
   – The 'John Ross', in dry dock after her trial trip, shows cracks at the attachments of the nozzle to the heel and the head box (Fig. 3.4).
   – This indicates that the problem is not just an incident, but structural. The nozzles are then removed, and the propeller blades are rounded off at the tips. The bollard pull without nozzle is reduced from 185 tons to 135 tons. I get the assignment to find a final solution as close as possible to the contracted bollard pull and free running speed.
   – 1979 This task is completed after three years when new nozzles are fitted.

2. Strength calculations

   I start my assignment with a search for a method to calculate the stresses at the crack locations. To my horror, I find that nozzles used to be built without any stress calculation. The plate thickness of the nozzle profile is simply derived from the hull plate thickness.

   Since the nozzle cracks are clearly due to fatigue (not shock), I begin with the development of my own fatigue stress calculation analogue to the Lips stress calculation for propeller blades. The nozzle thrust represents the load. The admissible stresses are taken from publications of Det Norske Veritas.

   It then transpired that the slotted nozzles of SAFMarine have excessive stresses. To be strong enough, the profile of the nozzles has to be almost massive. Since this is not feasible within the current manufacturing state-of-the-art, the slotted nozzle was discarded as a final solution.

3. Choice of nozzle and tank tests

   An open wheel on a tugboat is not acceptable because of the poor bollard pull. The room for a nozzle in the aperture was limited. The length of the slotted nozzle was only 0.3D (30% of the propeller diameter). A conventional 19A nozzle with a length of 0.5D would not fit into the aperture. After extensive search, I chose a length of 0.4D. Compared to a 19A

Fig. 3.4   Slotted nozzle of John Ross

nozzle, the profile has a longer flair and a shorter diffuser in view of the given location of the propeller disk. Both the shape and the length of the nozzle, therefore, make it far from conventional. Tests in the vacuum tank show that this DUC4 nozzle, as we called it, performs much better than the conventional 19A nozzle.

Comparison with the slotted nozzle:
- – Propeller diameter 5.0 metres versus 5.2 metres for the slotted nozzle.
- – Loss in free running speed 0.4 knots, or −2% compared to the slotted nozzle.
- – Bollard pull 195 tons, or +5.4% compared to the slotted nozzle.

The cost of tank testing amounted to almost hfl 100,000. Tests in bollard condition indicated that the attachments of the nozzle to the heel and head box had to be smoothened to avoid vortices and cavitation.

4. Commissioning

It is 1979 when the new nozzles are fitted at the shipyard James Brown & Hamer in Durban, where the 'John Ross' had been built. The nozzles were manufactured by Mützelfeld, Cuxhaven at a price of hfl 0.5 million. Kort Propulsion, UK was more than twice as expensive.

Trials for the 'John Ross' took place on May 2nd and 3rd. The nozzle for the 'Wolraad Woltemade' is then still on shore (Fig. 3.5). When the trials turn out to be successful, my assignment comes to an end.

On September 27th, 1981 I get the opportunity to inspect the nozzle of the 'John Ross' for a Lloyd's survey in dry dock in Vlissingen (Fig. 3.6). Everything looks fine, which gives me great satisfaction.

Both ships are still in service but have been sold by SAFMarine to Smit Global Group, owner of almost all large seagoing tugs on the planet, even in China. The name 'John Ross' has been changed into 'Smit Amanda'. Both ships now operate in South Africa under a National Pollution Prevention and Response contract.

Fig. 3.5 The DUC4 nozzle, Durban, April 1979, Vlissingen, September 27th, 1981

Fig. 3.6 'John Ross' in dry dock Vlissingen, September 27th, 1981

5. Epilogue

I fail to see myself as a pessimistic on-the-heavy-side designer. Vide the lightweight ultracentrifuges of URENCO, based on my design and patent (US patent 3,216,655 Nov. 9, 1965). On the road from invention to application in practice, neither pessimism nor optimism is appropriate. A critical realism about the circumstances and events which can be expected is required to achieve a good design. Thereafter, I look forward to the performance of my design with optimism and confidence.

I have to confess that without the failure of the slotted nozzle, my career would not have been half as fascinating. The spin-offs for further developments at Lips have been of great importance. The slotted nozzle forced us to develop our knowledge on nozzles to a much higher level than before. This enabled us to achieve a leading position, not only in nozzles for tugboats but also for dredging ships (trailers) and arctic icebreakers. I, therefore, doubt if my knowledge in the early seventies would have been sufficient to prevent the breakdown of the slotted nozzle of the 'Wolraad Woltemade'.

With hindsight, I see this case as the typical course of an industrial development. At some point a major deviation from the state-of-the-art is undertaken. In this case, a nozzle which is 40% shorter than applied before. Usual engineering rules of thumb are suddenly no longer adequate. New methods are then developed to overcome their shortcomings. The investment in expertise development Lips was forced to make because of the slotted nozzle has paid off more than satisfactorily.

Reply by van Gunsteren (2004):

Tactful wording is not a particular strength of mine, so give me a chance to rephrase. I see Jaap Wind as a brilliant, however conservative, mechanical designer who applies the ample safety margins which are required in a marine environment, and even more so when ice breaking, or dredging is involved.

His comments enable me to point out some relevant aspects:

1. In 1973, when Lips received the order for the two SAFMarine tugboats, Jaap Wind already had a worldwide reputation as a top-notch mechanical designer and I as a ship propulsion expert,[1] but neither of us was involved in the structural design of the nozzle for no other reason than the commercial consideration of limiting downside risk.

2. The step of reducing the nozzle length to 0.3D, which was 40% below the current state-of-the-art, was probably too big at that time. The involvement of Jaap Wind could have reduced that step to, say, 0.35D. The later wing nozzle has that length. Numerous wing nozzles have already been in service for years and none of them has failed. This proves that such a short nozzle with a slot in its profile can be made strong enough. The DUC4 nozzle was unconventional in the sense that its length was only 0.4D and its shape different from a conventional '19A' nozzle, but the idea was not new. Such a nozzle had already been proposed and tank tested before (van Gunsteren and Gibson 1974).

3. KaMeWa was the preferred supplier. Lips could only get the order because of the slotted nozzle. In other words: the slotted nozzle provided a competitive advantage. To keep that advantage, the strategically desirable course would have been to improve the invention by removing its shortcomings instead of discarding it in

---

[1] For instance, the inventor was the spokesperson of all European propeller manufacturers in the international normalisation ISO-TC8 and in Lloyd's Panel for Ship Propulsion.

favour of a conventional nozzle without a slot. Admittedly, this would require side struts which bring along their own risks. They have to be carefully positioned: connection to the nozzle sufficiently behind the leading edge and attachment to the hull not too close to the nozzle. To get an innovation off the ground requires a strong will to do so after my departure that attitude was no longer prevailing.

The ironic conclusion is that the cause of the slotted nozzle failure is no more and no less the incompetence of the sales director who committed the blunder of leaving superior in-house expertise deliberately unutilised.

The Fear of Innovation

4.1 The Ambivalence to Innovation

Technical progress, generated by numerous innovations in relevant domains of our economy, has brought our Western society unprecedented prosperity and wealth. The interest of society at large, therefore, is that the stream of complementary innovations will continue also in the future. The relevance of innovation for our quality of life and even more our quality-of-work-life is undisputed. As a corollary, innovation is praised as a blessing, rewards are issued for innovating accomplishments and innovation-earmarked subsidies are made available, in particular for "backward" sectors of the economy. Criticism on the widely prevailing point of view that innovation is good for everybody seems to be an undiscussable issue, a taboo.

Nevertheless, we can observe in practice that strong forces unavoidably obstruct the innovation process. Why?

The cause is twofold: (1) Shift of power and (2) Dynamic conservatism.

4.2 Shift of Power

Innovation changes the balance of power within an organisation. Machiavelli already taught us in the fifteenth century (The Prince):

> There is nothing more difficult to take in hand, or more perilous to conduct, or more uncertain in its success, than to take the lead in the introduction of a new order of things because the innovator has for enemies all those who have done well under the old conditions, and only lukewarm defenders in those who may do well under the new.

L. A. van Gunsteren, A. G. Vlas, *The License Giver Business Concept of Technological Innovation*, Future of Business and Finance, https://doi.org/10.1007/978-3-030-91123-2_4

Example: Fear of Creative Output

Typical idea killers encountered in practice are the following:

• Ever tried by someone else?	• Too theoretical
• Outside our scope of business	• Not practical
• Against company policy	• We tried that 5 years ago
• Does not work in our trade	• Damages our current business
• We do not have enough money	• Board of Directors would not like it
• Collect some more information	• *Idea killer no. 1: Silence*

Idea killer no. 1—to ignore completely the proposed idea—was once applied in what seemed to be a most valuable innovative effort. The traditional way of dredging, namely from a ship which is kept at its location by spuds or anchor lines, becomes problematic in very deep water and in exposed areas. An obvious solution is not to dredge from a ship but instead from a vehicle on the sea bottom. A wealth of ideas related to this was generated by a multi-disciplinary group. In addition to some key scientists from the firm's own R&D department, some carefully selected experts from the outside participated in the exercise: a biologist, a physicist, an army general (dredging can be seen as attacking soil), a specialist in farm equipment, and an industrial designer. The final report, containing a dozen new concepts along with some 25 inventions, was completely ignored by the traditional dredging people. Why? Because all those ideas were simply too threatening to them as can be illustrated by the following hypothetical dialogue:

- Dredging man: 'What should we do with our current fleet when that new way of dredging becomes operational?'
- R&D-man: 'By the time that we are pretty sure that the new technology works we'll sell our ships to a competitor who is then later at an extra disadvantage'.
- Dredging man: 'What should we do with people (like me) having expertise in ships but not in underwater technology?'
- R&D-man: 'We'll sell them with the ships of course…!'

4.3   Dynamic Conservatism

Donald Schön (1971) has coined the term *rational view of innovation* to describe the belief that innovation is a manageable process in which risks are controlled by justification and review. His assertion is that the rational view of innovation ignores or violates actual experience and is, therefore, a myth.

The experience of the authors confirms this point of view. Table 4.1 shows what happened with some inventions of the first author (van Gunsteren 2004).

Lesson from these failures: Beware of limiting downside risk; it can kill the innovation.

Clearly, what happened is far from an orderly goal-oriented process of decision-making and optimisation. These inventions could not have been predicted beforehand.

Table 4.1 What happened with some of van Gunsteren's inventions

Invention	Cause of failure
Control system for the steering of ships (patent granted March 18, 1966)	• No patent application in USA. • Launching customer limiting his downside risk by a switch for returning to conventional steering.
Ventilated controllable pitch propeller	Complementary military invention was abandoned.
Special ring propeller	Offending launching customer by sales manager who insisted on an excessive price.
Slotted nozzle	Sales director's decision to limit his downside risk by entrusting the construction design to the manufacturer who could be held responsible in case of failure.
Shafting arrangement for contra rotating propellers (CRP)	Decreasing oil prices reduced the interest in fuel saving innovations; the 16% fuel saving of CRP was no longer considered enough to justify the investment.
Blade shape of oars	Termination of paying taxes in view of limited commercial prospects.
Wing nozzle	Initial failure due to launching customer limiting his downside risk by preserving his right to remove the nozzle from the ship. Later the wing nozzle became a success thanks to a second launching customer.
Steering arrangement for race rowing boats (patent granted September 20, 2019)	Prototype testing promising so far; development still going on.

Innovation creates a situation of uncertainty, which should not be confused with risk.

Risk is a possible future loss or damage to which occurrence a probability can be assigned. Risk allows mathematical modelling that enables to get valuable insights from computer simulations.

Uncertainty cannot be represented in a mathematical model. A situation is uncertain when it requires action but resists analysis of risks. The need for action is clear but not at all what to do about it. There are no clear objectives to reach, no measures of accomplishment and no clarity on what to try to control.

Case: Covid19-Crisis

After the outbreak in December 2019 the need for governmental interference was evident but not at all what to do about it. Precious weeks were lost before governments decided to act with far-reaching measures. Only Taiwan had learned its lesson from the SARS outbreak of 2003.

The feeling of uncertainty is anguish. To avoid the anxiety caused by the inherent uncertainty of disrupting innovations, organisations display *dynamic conservatism*, the tendency of organisations to fight like hell to remain the same.

As a result, the path from an initial idea to ultimately contributing to technical progress is paved with roadblocks that are difficult to overcome (Fig. 4.1).

Required   **Phase**

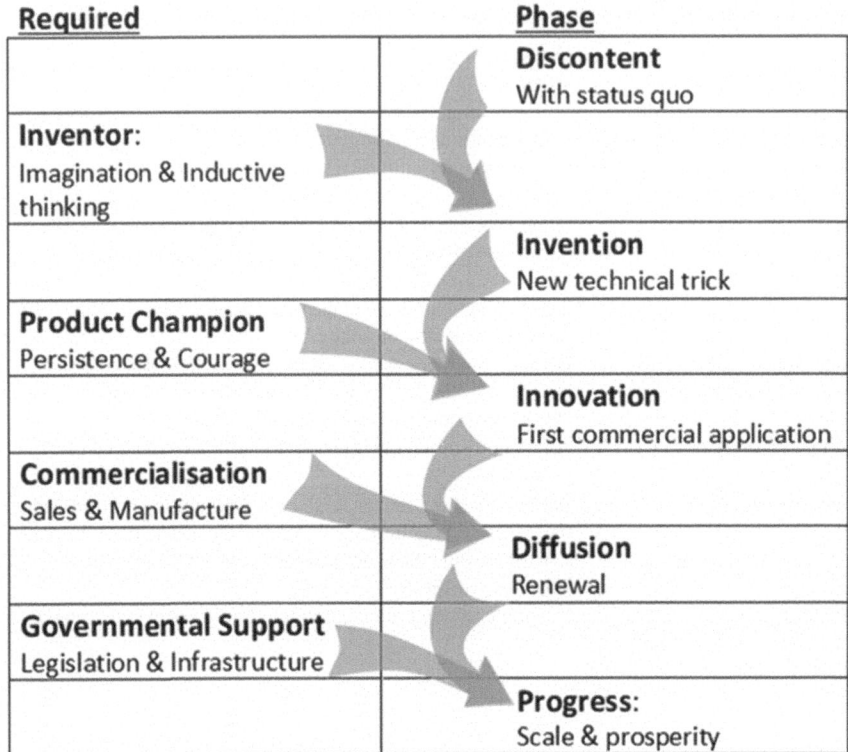

Required	Phase
	Discontent With status quo
Inventor: Imagination & Inductive thinking	
	Invention New technical trick
Product Champion Persistence & Courage	
	Innovation First commercial application
Commercialisation Sales & Manufacture	
	Diffusion Renewal
Governmental Support Legislation & Infrastructure	
	Progress: Scale & prosperity

Fig. 4.1 Stages of the process of innovation

To steer a promising idea around many potential obstructions requires persistence and creativity throughout every phase at various transition points.

One such critical transition point is finding a suitable launching customer. In all four 'nozzle-related' cases, ring propeller, slotted nozzle and wing nozzle twice, the inventor had to actively involve himself to break through stalemate situations.

Case: Control System for the Steering of Ships
The history of van Gunsteren's first patented invention is described by him as follows.

The steering characteristics of a ship are of great importance when a collision is to be avoided by putting the rudder hard on board. Most ships have their rudder angles limited to 30–35°. If the rudder would be put at larger angles, stalling (flow separation) would occur and the rudder would only brake, but not generate much side force. When the ship begins to turn, usually around a point at about 10% of the length of the ship from the bow, a transverse velocity component to the rudder reduces the angle of attack. To compensate for this effect, the rudder must be put at a higher rudder angle if the side force of the rudder is to be maintained.

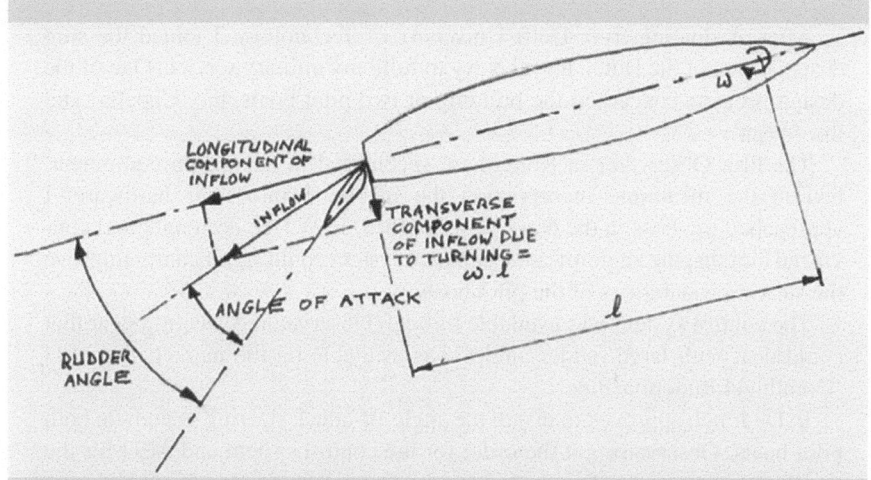

Fig. 4.2 Steering on angle of attack

This phenomenon is well known to anyone who has sailed in a dinghy. When going through the wind, the rudder is at first put at a moderate rudder angle. When the dinghy begins to turn, the rudder is moved to a larger rudder angle to maintain the rate of turning of the dinghy.

This procedure can be programmed into a steering control system (Fig. 4.2).

The signal from the bridge is no longer the desired rudder angle, the angle between the ship and the rudder, but the desired angle of attack, the angle between the incoming flow and the rudder. The control system automatically corrects the rudder angle for the transverse velocity component caused by the turning of the ship. The rotation of the ship at any moment is determined by the recent history of the rudder angle, the rudder angle as a function of time in the period prior to the present moment. No other input signal is required than the rudder angle itself.

A second possibility is specified in the patent application, namely that the rotation of the ship is measured with a gyrocompass. This version avoids the necessity to introduce a correction factor for the loading condition of the ship.

I asked my father to cover the costs of the patent application. After explaining that as a student of technology it would be useful to get some experience with patents, I presented two options to him:
1. Patent application only in the strictest country in regard to granting patents, which happened to be The Netherlands. In that case the cost of applying had to be considered 'sunk cost'. The patent would be mine.
2. Patent application in the leading maritime countries. The considerable cost as well as any future revenues would be for my father.

Since he chose the first option, I filed the application only in The Netherlands (Dutch patent application no 66.03583, 'Control system for the steering of ships', patent granted March 18, 1966).

After graduating from Delft University of Technology, I joined the ship design office of the Dutch Royal Navy to fulfil my military service. One of the design projects concerned the building of two pilot boats, the 'Capella' and the 'Wega'.

The firm Observator in Rotterdam, specialised in navigation equipment, had in the meantime incorporated the invention into their hardware. I approached my boss in the Navy design office, Ir. W.P.H. de Jongh, and convinced him that the angle-of-attack control system could significantly improve the steering capabilities of the pilot boats.

The control system was available through Observator. A steering gear that could deal with large rudder angles was available on the market: the AEG 'Drehflügel Rudermachine'.

Ir. De Jongh approved to install the angle-of-attack steering system on both pilot boats. Observator got the order for the control system and AEG for the steering gear (in spite of protests from the Dutch firm Stork-Jaffa, which was unable to offer a steering gear that could handle large rudder angles).

The trial trips were an undisputed success. The new steering system reduced the turning circle diameter from 130 to about 100 m. This result could easily be established, for Ir. de Jongh had demanded that by means of a switch on the bridge the captain could change to conventional steering at any time.

A few months later, I had to appear before the Deputy Minister of Defense together with two other reserve officers, one of the Royal Air Force and one of the Royal Army, who both had also produced an invention during their military service. We were allowed to keep our patents. In the future, however, inventions patented and developed during military service would be property of the Ministry of Defense.

I then got a lesson in patent exploitation. An American firm (Westinghouse) was interested in buying the patent, but no longer showed any interest when they learned that no patent application had been filed in the USA.

Some years later, one of the pilot boats had a collision. The investigator of that accident gave me a phone call to convey his conclusion to me. If the switch on the bridge had been set to steering on angle-of-attack, the collision could have been prevented. It turned out that only the first captain, who had been briefed by me, had used the angle-of-attack steering system. His successors had felt themselves uneasy with a steering system unfamiliar to them and had always put the switch on conventional steering.

This reminded me that I had expressed my reservations about that switch to Ir. de Jongh, but he had given me no choice: without that switch, no full-scale application of the invention. In doing so, Ir. de Jong limited his downside risk, but at the same time the possible benefits of the innovation. Just as my father did, when he opted for patent application in only one country.

Could the collision of the iceberg with the Titanic, 'the ship that God himself couldn't sink', have been prevented if the turning circle diameter was 23% less by means of the angle-of-attack control system? We will never know.

Dynamic conservatism can take many forms. The most common and most serious one is the tendency to defend the current business against any disrupting innovation instead of embracing it before it is too late.

4.4   The Story of the Ring Propeller

The story of the Ring Propeller is told by van Gunsteren as follows.

A ring propeller is, as the word indicates, a propeller with a ring attached to the blade tips. This ring rotates with the propeller. The ring has an aerofoil shape similar to the nozzle of a ducted propeller (Kort nozzle). The ring propeller resembles this type of propulsion in many aspects. The main difference is that the ring rotates with the propeller. The consequent viscous forces on the rotating ring produce an extra torque. As a result, the optimum rotational speed (rpm) at a given diameter is considerably lower than in the case of an open wheel or non-rotating duct.

The ring propeller has already been applied long ago. Figure 4.3 shows a photograph of a ring propeller on a twin screw barge on the river Yang Tse Kiang in China in 1927.

Such prior art means that a strong patent on the principle of a ring propeller is no longer possible. Patent applications on construction details would be possible but can always be circumvented. In such cases, development has to be left to big players in the market who need patent protection much less than small companies.

Fig. 4.3 Ring propeller as mounted on a twin screw barge in China in 1927

Fig. 4.4 Launching
customer implementation

Since the ring propeller seemed to be an attractive option in cases where a ducted propeller would be favourable but impossible because of problems with the attachment to the hull, I investigated its potential quite extensively during the late sixties (van Gunsteren 1970, 1971).

My efforts to get the ring propeller accepted followed the same pattern as the slotted nozzle. The development started with conducting the systematic open water tests needed to design a propeller with the right pitch.

Simultaneously, the process was developed to manufacture the ring from reinforced plastic.

Subsequently, we found a launching customer who was prepared to fit a ring propeller to one of his coasters (Fig. 4.4).

The hydrodynamic performance turned out to be better than expected. The ship with the ring propeller outran the two sister ships which had been in dry dock for hull cleaning at the same time.

After the ring propeller had been in service for half a year, a similar incident took place as later happened with the slotted nozzle. A heavy pounding was noticed during manoeuvring close to the port of Lisbon. Upon investigation it was found that the ring had disappeared. The propeller was badly damaged (see Figs. 4.5 and 4.6 for original design of connection between ring and propeller). From the damage could be concluded that a heavy object must have struck the propeller. Although a conventional propeller would also have had severe damage under such circumstances, we decided to strengthen the connection between the propeller and the ring.

The new attachment, designed by Jaap Wind who had not been involved until then, looked very sturdy indeed (see Fig. 4.7).

A new ring propeller was manufactured using Jaap's sturdy attachment construction. Then something unexpected happened: our sales manager decided that our

Fig. 4.5 Damaged
propeller

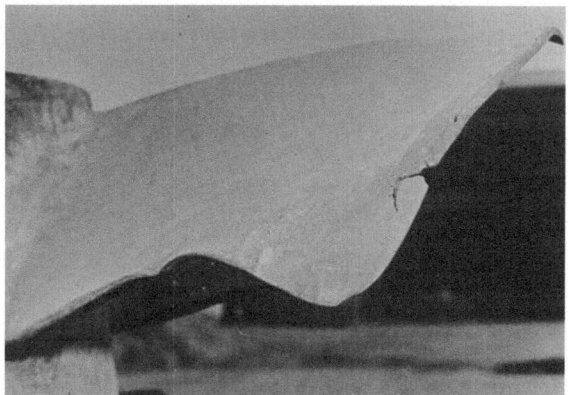

Fig. 4.6 Drawing of
original connection

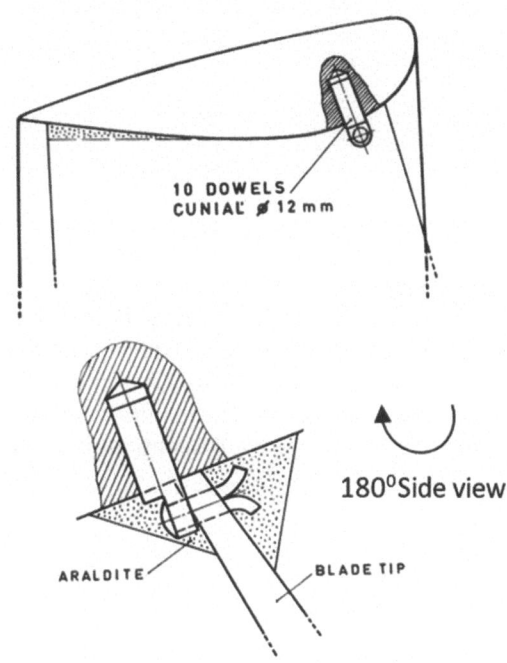

launching customer had to pay twice the price of the first ring propeller, which had
been sold at the price of an equivalent open wheel.

Our launching customer was furious and refused to fit the new ring propeller to
the ship. Of course, I protested vehemently against the outrageous pricing by our
sales manager, but he did not give in. He argued that, since the development was
completed, the pricing was now his prerogative and that he wished a price for the
ring propeller that would allow the same profit margin as a conventional one. I
pointed out that pricing a new product with profit margins of a cartel situation would
kill any innovation. It then dawned upon me that this was exactly what the sales
manager was after.

Fig. 4.7 Drawing of new
strengthened connection

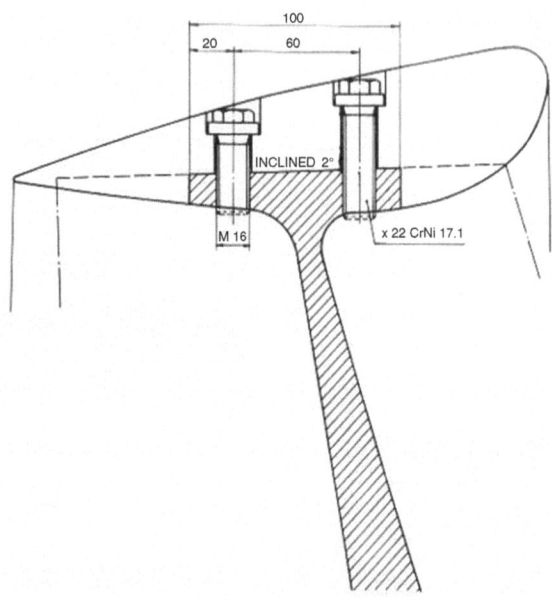

The stability of a cartel, in essence a form of a monopoly, can be disturbed when one of its members gets an advantage over the other members by a patented, or hard to copy, invention.

That was the end of the ring propeller innovation. I could have revived the matter a few years later when I had more power in the company, but I decided that the slotted nozzle was a better cause to fight for. It offers more or less the same advantages as the ring propeller without the disadvantage of an about 30% lower optimum rotational speed compared to a conventional shrouded propeller.

The Not-Invented-Here-Syndrome

5

5.1 The Inventor's Fallacy: Overrating the Invention's Merits

Inventors tend to overrate the technical merits of their inventions and underestimate the marketing efforts required for market acceptance. Valuable inventions often fail commercially as a result of counterproductive behaviour from the part of the inventor, such as

- Demanding a too high license fee.
- Pursuing a monopoly.
- Refusal to share benefits with partners taking care of marketing.

Sometimes, the inventor has first to retire before successful market launching becomes possible. In other cases, the innovation is accepted by the market but the inventor, or his organisation, is not in a position to reap an appropriate financial reward.

5.2 The License Fee

When License Giver and License Taker are completely separate companies a license fee has to be agreed upon. But what is an appropriate license fee? If the license fee is too high, the License Taker prices himself out of the market and if the license fee is too low, the License Giver cannot recover his RD&D costs. The issue is often aggravated by the inclination of inventors to underestimate the efforts needed from the License Taker to gain acceptance of the new product in the market. License fees can vary widely, because the advantages of the new products can be very different. Discussion on license fees is pointless as long as the financial benefits of the new product over the existing product have not been established. Let us call these

L. A. van Gunsteren, A. G. Vlas, *The License Giver Business Concept of Technological Innovation*, Future of Business and Finance,
https://doi.org/10.1007/978-3-030-91123-2_5

benefits *the pie* that has to be divided between the License Giver and the License Taker. We recommend the following rule of thumb for this:

The license fee to be paid by the License Taker to the License Giver should be 30–50% of the 'pie' (benefits of the new product), dependent on the engineering support provided by the License Giver.

Example: Licensing New Type of Boots to a US Manufacturer

A European (License Giver type) boots manufacturer wishing to license the manufacture of a new type of boots to an American manufacturer asked advice on how to negotiate the license fee. The question on how much 'pie' there was to divide was easily answered in this case: the alternative for License Taker was to import the boots from Europe in which case 15% import duties had to be paid. So, the 'pie' was 15% of the turnover concerned. People dislike being confronted with just one possibility. Therefore, three options were offered:

1. A fee of 5% and all engineering support to be paid extra, both man-hours and expenses.
2. A fee of 7%, engineering support man-hours free of charge but expenses (lodging and travel) paid extra.
3. A fee of 8% that includes all required engineering support.

The License Taker chose the first option (5%), as this was the best for him on the longer term, and the license agreement was signed. The supervisory board asked the managing director of the License Giver how he had achieved a license fee of 5%, whereas nowhere else in the industry more than 2% had ever been realised. The answer was 'by first agreeing on the size of the 'pie' and then on a fair division of it'.

The new product (or process) always provides financial benefits compared to the current one. Usually, engineers of both parties have no difficulty to agree on how the financial benefits work out, be it reduced manufacturing cost, better performance, lower maintenance cost, etc. Once the size of the 'pie' has been established, negotiations about the fee become a lot easier.

This is in particular important in the case of a small innovation company (License Giver) negotiating with a big License Taker company interested in marketing the new product worldwide. If the size of the pie and its fair division are not separated from each other, professional negotiators from big firms can easily outsmart small companies or pressure them into an agreement. The small company, often the inventor, should establish its walkout price at 30% of the 'pie' and stick to it.

Relating the license fee to the turnover of the License Taker, as is usually done, has the advantage that market price increases and inflation is automatically accounted for. The disadvantage is that the License Taker has to reveal his prices. It is sometimes better to relate the license fee to a physical parameter, for instance, the diameter of the rotor of a windmill or the diameter of the rotor of a pump and correct for inflation by a generally accepted index.

The rule of thumb—30–50% of the 'pie' to the License Giver, depending on the engineering support included—implies that for marginal innovations (small pie) engineering support should be offered free of charge in order to achieve a license fee worth bothering.

5.3   The Story of the Wing Nozzle

Three years after the slotted nozzle patent rights were transferred to the inventor, he founded the innovation company Van Gunsteren & Gelling Marine Propulsion Development BV for the further development of the slotted nozzle. The four share-holders, having equal shares, were two former colleagues from his time at Lips (Ab Brink and Pierre Pellenaars), himself and Jaap Gelling, who had just graduated from the Faculty of Naval Architecture of Delft University of Technology. As part of his studies, he had successfully managed the development of a rowing eight of aramid fibre, an admirable achievement and a convincing proof that he would be the right partner.

The history of this partnership is described by van Gunsteren (2004) in this section.

Since we suspected that vibrations had been a major cause of the failure of the nozzle of the 'Wolraad Woltemade', we first developed software to analyse vibratory behaviour of nozzles. The computer programme enabled us to compute the natural frequencies of any nozzle configuration. Calculations on the slotted nozzle of the 'Wolraad Woltemade' revealed that resonance had been the main cause of the break down and also that fitting struts would have been an effective remedy.

Another point of concern was cavitation coming from the slot that could cause cavitation erosion on the propeller blades. This brought us onto the idea of locating the slot behind the propeller instead of in front of it. This configuration, initially called a *flapped nozzle*, is nowadays known as Wing Nozzle.

Figure 5.1 shows the Wing Nozzle in the aperture of a coaster.

Systematic open water tests on the wing nozzle, conducted at MARIN proved that the same advantages could be achieved as with the slotted nozzle.

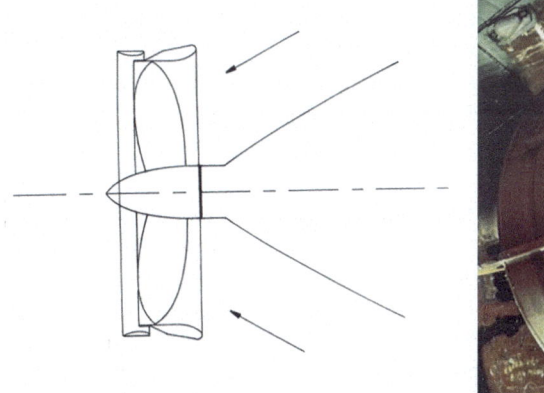

Fig. 5.1 Wing Nozzle in the aperture of a coaster

The wing nozzle solved another problem for us: the expiration of the slotted nozzle patent. A patent on the wing nozzle was granted, in spite of the comment of the examiner that it was a variation of my patent from 1969.

Supported by reliable software, thoroughly tested by model experiments, and protected by our new patent, we were ready for a launching customer.

Finding an appropriate launching customer is critical in any innovation process and the wing nozzle was no exception. Since nobody wants to be a guinea pig, the launching customer must have more reason to go ahead than only the benefits of the invention. In our case that reason was that I had, as a management consultant, helped to save the shipping company Nigoco from bankruptcy. Its managing director owed me a favour. The favour was a contract for wing nozzles on a series of six new buildings, known as COB vessels, to be built by the shipyard Van der Giessen-de Noord.

Like in the cases described before, the managing director of Nigoco wished to limit his downside risk. A clause was included in the contract giving him the right to remove the nozzle from the ship for whatever reason.

The yard was not at all pleased with the demand of the ship owner to fit wing nozzles on the COB vessels. The tiny profit margin on their contract could easily evaporate by any delay due to the complication of the wing nozzle.

The non-cooperation of the yard, if not outright obstruction, included the following:

1. The support construction for the nozzle was not only priced at a very high unit rate, but was also far too heavy. After interference from our part, approval from the classification society was obtained for a construction that was three times lighter than initial proposition from the yard.
2. Since their contract with the owner did not include the usual clause of a minimum required trial speed, the yard had abstained from any model testing. We considered model testing essential and offered to pay one third of the cost and proposed that the other two thirds would be shared by the owner and the yard. The owner agreed but the yard refused to contribute anything. Ultimately, we shared the cost of the model tests with the owner.
3. The discussion on who should pay for the model test had taken so much time, that any modifications in the nozzle design based on the test results would no longer be possible without extension of the delivery time. Knowing that the yard could not keep their own schedule anyway, we asked for an extension of a few weeks. The request was bluntly refused by the yard. After we delivered the nozzle according to schedule, it laid idle on the premises of the yard for almost 2 months. The requested extension of the delivery time, which could have been granted without any disadvantage for the yard, would have enabled us to adapt the nozzle design on the basis of the test results. These results were disastrous. MARIN strongly disapproved the blunt lines of the aft body of the ship and predicted that flow separation would occur, causing extra resistance and extremely poor inflow into the propeller disk. But design changes were out of the question at that stage.

4. At the trial trip, the ship showed severe vibrations on the bridge, for which the nozzle was blamed. The yard left the owner no other option than to give permission to remove the nozzle from the ship. This was done, without notice to us, by cutting the nozzle into pieces making it impossible to preserve it for use on other ships with a propeller diameter of 3.5 m.

Refusing to allow a few weeks delay in delivering the nozzle to enable us to incorporate the model test results in the design is undoubtedly the most serious of these points. The yard's aft body design was so bad that changes in the nozzle design might have been still insufficient, but they would at least have alleviated the problem.

In the meantime, we had secured the order from Damen Shipyards for wing nozzles on a series of five coasters. When the word spread about the removal of the wing nozzle at Van der Giessen-de Noord, voices within Damen Shipyards became louder to get their order cancelled. The discussions only came to a halt after interference of Kommer Damen himself who decided: 'The only way to find out if that nozzle really works is to try it out on one of our ships'.

At the trial trip, it soon became apparent that the nozzle performed as predicted and also reduced noise in the quarters in the aft ship from above to below the legally allowed level.

The innovation was saved.

The Damen Group then took over the intellectual property rights, the computer programmes and, most importantly, the engineer who had been involved in the entire development process: Jaap Gelling. Apart from the development of a shaft arrangement for contra-rotating propellers which we were involved in, Van Gunsteren & Gelling Marine Propulsion Development BV became a dormant company.

Figure 5.2 shows a comparison of the profiles of the Conventional Nozzle 19A, the Slotted Nozzle and the Wing Nozzle.

Fig. 5.2 Comparison of the profiles of the Conventional Nozzle 19A, the Slotted Nozzle, and the Wing Nozzle

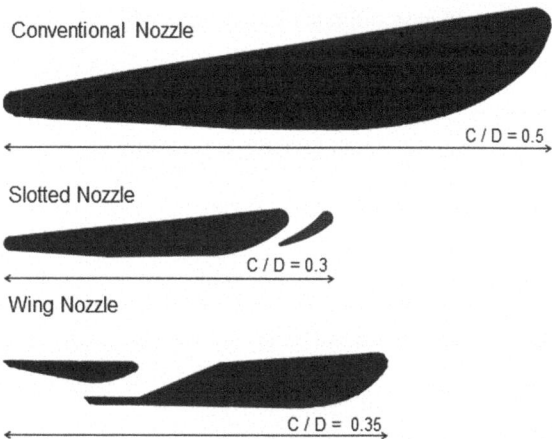

Conventional Nozzle

C / D = 0.5

Slotted Nozzle

C / D = 0.3

Wing Nozzle

C / D = 0.35

Nowadays, hundreds of ships with a Wing Nozzle are sailing around the globe earning money for their owners. From this point of view the Wing Nozzle innovation can be regarded to be a success.

The diffusion of the innovation, however, was considerably slowed down by the short-sightedness of the director of the operating company entrusted with the exploitation of the Wing Nozzle innovation.

Both strategic mistakes that are common when it comes to exploiting a well-patented innovation were made:

1. Aiming at a monopoly; the director systematically refused to grant a license to a competitor.
2. Setting an extra profit margin on the price of a Wing Nozzle.

These policies were maintained until retirement of the director.

It took this long to become evident that these policies are counterproductive because of the director's unwillingness to pay the contracted license fee of 6% to the inventor. Paying a licence fee implies recognition that someone has been smarter than you. This generates feelings of jealousy and false pride, a phenomenon known as the *Not Invented Here Syndrome (NIHS)*.

The managing director let his NIHS prevail instead of embracing the License Giver business concept.

The wing nozzle has been invented by van Gunsteren en Gelling. They published about it under the name Flapped Nozzle. Shortly after acquiring the patent rights, the director renamed the nozzle Wing Nozzle to conceal its origin. The Wing Nozzle is presented in the firm's documentation as an *in-house development*. This is clearly not true, as can be inferred from the patent description.

The License Giver business concept aims at uniqueness by *design leadership*, not through a monopoly or cartel. Everybody is allowed to apply the Licence Giver's design provided an appropriate license fee, in whatever form, is paid to the License Giver.

Instead of refusing to grant a license to a competitor, the wing nozzle manufacturer should have offered a license fee of 8%. This would have been to the benefit of everybody involved:

1. The manufacturer himself would earn 2% on sales of his competitor.
2. The inventor would earn 6% on sales by the competitor.
3. The competitor would promote the wing nozzle in his dealings with customers instead of opposing its application.

The latter has great impact on the diffusion of the innovation.

The inventor has to cash-in before the patent expires. The delay in diffusion of the innovation has significantly limited the return on the inventor's investment.

When selling a patent, the inventor should be aware that the Not-Invented-Here-Syndrome can spoil any promising deal.

The Rise and Decline of Innovative Capability

6

6.1 Challenges in Various Phases of Corporate Development

We can distinguish three different phases in the history of a corporation, each having their typical challenges (Table 6.1).

In the pioneer phase, there is little room for innovating endeavours. Failure could easily draw the company into bankruptcy.

In the expansion phase, the firm has accumulated sufficient financial buffer to afford the failures that are inherent to innovative development projects. Family enterprises being still in their expansion phase with the family in control and located in a cluster of their trade constitute the natural candidates for innovating efforts: the watch industry in Switzerland and the early automotive industry clusters that emerged in Germany, the UK, USA and other countries. Somewhat comparable to the earlier-mentioned industrial clusters, the German 'Mittelstand', a special Small & Medium Enterprise business environment, has also proven to be a resilient and very fertile breeding ground for innovative efforts by family enterprises.

Often one can notice a change in top management in the transition from the expansion phase into the maturity phase. The CEO with deeply rooted affinity with the firm's products and core technology is succeeded by someone with a background in finance or politics.

At the same time the company experiences ongoing pressure to keep up growth rates, as required by an increasing diverse and less-connected base of investors. The drive to grow often points to diversification into more drastically new markets or even technologies. Diversification and changing leadership can easily become a recipe for organisational confusion, lacking the strategic clarity and driven leadership prevailing in the expansion phase.

© The Author(s), under exclusive license to Springer Nature Switzerland AG 2022
L. A. van Gunsteren, A. G. Vlas, *The License Giver Business Concept of Technological Innovation*, Future of Business and Finance,
https://doi.org/10.1007/978-3-030-91123-2_6

Table 6.1 Challenges in various phases of corporate development

Phase	Challenge
Pioneer phase	Getting critical mass
Expansion phase	Recruiting and retaining high-flyers
Maturity phase	Fighting bureaucracy

A token that bureaucracy has sneaked in, indicating that the company has entered the maturity phase, is that discussions in the organisation are predominantly about process—rules, regulations, and policies- and no longer about content.

It seems to be inevitable that sooner or later the temptation can no longer be resisted of defending the current profitable business instead of focusing on the next generation of technology.

The usual way to cope with these size-related issues is to adopt the multi-divisional structure: splitting up the organisation into semi-autonomic units being allocated the responsibility of maintaining market share and profit in a single business or market. The multi-divisional structure has become the norm for most large organisations. In fact, the concept became elevated to an unquestioned doctrine that has prevailed from before WWII until the early nineties.

Unfortunately, the multi-divisional structure is not particularly conducive to innovation as is explained in the next section.

6.2   The License Giver's Need of a Functional Organisation Structure

How should the innovative organisation be structured?

The paradoxical answer to this question is: as little as possible. To innovate means to break away from established patterns. The innovative organisation minimises all forms of standardisation for coordination. Ideas should be allowed to flow up and down and to easily cross functional boundaries. Innovative organisations have a flat structure and blurring functional boundaries. Innovative organisations are of an integrative nature (Moss Kanter 1983), as opposed to a segmental one. They extensively make use of task forces to overcome functional boundaries.

Why it is so important to avoid the most common and most serious error in organisations: having too many hierarchical levels?

Because every level introduces:

1. A filter of information (Table 6.2).
2. A barrier that has to be overcome to get ideas accepted which originate from lower levels in the organisation, summarised in the following rhyme:

along this tree
from root to crown
ideas flow up
and vetoes down

Table 6.2 Each hierarchical level introduces a filter of information

• Director tells all he knows = 100%
• Manager recalls 80%
Tells all he knows
• Project manager recalls 80%
Tells all he knows
• R&D worker recalls 80%
Director wonders why the R&D workers are so dumb: $0.8 \times 0.8 \times 0.8 \times 0.8 = 0.4096 = 41\%$

Table 6.2 shows that a 20% loss of information at each of the four hierarchical interfaces results in only 40% of the message arriving at where the action should take place. Bottom up the situation is usually even worse as subordinates tend to limit messages to their bosses to what they expect they like. With, say, a 30% information loss at each interface only 20% of the original information will ultimately reach the highest-ranked manager.

An effective measure to avoid ideas from a lower hierarchical level being blocked is to make it normal for employees to have regular contacts with both their manager and the manager above him or her. This can be done by defining certain matters that must be directly submitted to the manager of someone's supervisor. When the subordinate expects that his supervisor may reject his idea for whatever reason, he can define it into the category for the higher manager without being blamed for bypassing his supervisor.

Overcoming functional boundaries between departments is equally important. Innovation is always of a multi-disciplinary nature. When a new concept is put into operation numerous teething problems have to be solved. This requires close and smooth cooperation between individuals and departments covering different disciplines.

In the innovative organisation, groups of people work closely together towards the achievement of a common goal, as the situation requires fairly independently from hierarchy and functional boundaries. Following Alvin Toffler who introduced the term 'Adhocracy', Henry Mintzberg (Mintzberg 1979) has described its features in considerable detail (Table 6.3).

In an Adhocracy type organisation actual power is not necessarily vested in the highest placed executive. Power shifts depending on the problem at hand. The expert who has particular expertise on a particular subject is tacitly allowed to take the lead. At other stages, when other skills are required, someone else will play the leading role. The formal structure of the innovative organisation must be sufficiently loose to allow the shifting of power when required by the situation. An equally important prerequisite is a feeling for the dynamics of innovation is well represented at the top of the organisation.

The Adhocracy is not embraced by everybody. Many people, in particular creative ones, dislike structural rigidity and concentration of power. The adhocratic organisation is organic and decentralised, making it for them a great place to work in.

Table 6.3 The Adhocracy organisation features

• Highly organic structure
• Little formalisation of behaviour
• Small market-based teams
• Key coordinating mechanism:
• Mutual adjustment, within and between teams
• Selective decentralisation to and within teams, located at various places, involve mixtures of line managers, staff and operating experts
To innovative means to break away from established patterns, so the innovative organisation cannot rely on any form of standardisation for coordination A. Todfler & H. Mintzberg

But others prefer a life in a bureaucratised setting characterised by stability and well-defined relationships. When the latter gets dominance over the former, the innovating capability of the organisation will inevitably decline.

The License Giver business concept serves to avoid this to happen by keeping the two groups—those preferring *challenge* and those preferring *stability*—separated in the License Giver and License Taker organisations. In this way cooperation between the two groups becomes possible without an internal power struggle.

Case: Risk Aversion of CFO Lips Propeller Works

With the delivery of the 35,000 HP CPP installations for the Seatrain container vessels in the early seventies, Lips had successfully achieved the transition from License Taker to License Giver. But the company's CFO still displayed the risk aversion of a License Taker, as became painfully apparent in the following incident, reported here by van Gunsteren.

The US Navy DD963 project in the early seventies concerned 30 destroyers to be built by Litton industries. For the 60 CPP installations, only KaMeWa and Lips were invited to bid. For the execution in the US, Lips had secured assistance from Pratt & Witney. The situation was quite similar to the slotted nozzle: KaMeWa had an excellent track record, Lips was particularly qualified in suppressing noise radiation due to cavitation. My technical presentation for the purchasing committee focussed on the weakest points of our CPP-design enabling to assess shock resistance in relation to other systems on board. This approach of focussing on the weak points of our design instead of emphasising its strong points as one would expect, was highly appreciated the chairman of the purchasing committee told me later.

In the meantime, KaMeWa USA had managed that during the bidding process not a single document would be allowed to leave the country. As I knew this before returning to the Netherlands, I memorised all relevant data. Of course, our bidding price needed approval from our CFO. The request for proposal specified that during a few decades any necessary repairs above ordinary preventive maintenance would be for the manufacturer's account. This item was a substantial part of the cost price. I proposed to completely ignore this clause because the probability of it ever becoming relevant was very low for two reasons: (1) The proven reliability of our design and (2) The conduct to be expected from the captains of the ships. When in peace time, no commander will ever operate the ship at full power and during war time nobody will care about what is specified in the contract. But our CFO insisted nevertheless to keep these unnecessary costs in our bid. When all pros and cons are balanced the bidding price usually becomes decisive for awarding the order. Apparently, our CFO did not want our bid to be successful. A key success factor for the License Giver is the realisation of getting the largest possible number of applications deployed globally. To leave deliberately such a huge order to the competition is an unforgivable mistake for a company with a License Giver ambition.

What will happen if the License Giver follows the multi-divisional doctrine instead of adopting the Adhocracy form of organisation?

To answer this question, let us return to the stories of the ring propeller (Sect. 4.4) and the slotted nozzle (Sect. 3.3).

In line with the multi-divisional doctrine, Lips adopted in the early seventies a profit centre concept that granted considerable decision power to the managers of the profit centres. Accordingly, the manager of the profit centre Monobloc Propellers was authorised to kill the ring propeller innovation that would have a short-term negative impact on the profitability of his profit centre.

The manager of the Controllable Pitch Propeller profit centre had earned his position by becoming a son-in law of Max Lips (and losing the position when the marriage was over). He had thereby reached his level of incompetence as became apparent from the blunder of entrusting the strength calculations of the slotted nozzle to Kort Propulsion Ltd. instead of involving the available brilliant in-house expert, Jaap Wind. He let commercial considerations prevail over prioritising design quality, a deadly sin for a License Giver company. The profit centre concept, however, empowered him to take the decision that ultimately killed the innovation.

6.2.1 Experts-Leading-Experts

The multi-divisional doctrine entails that when firms grow large, they must shift from a functional to a multi-divisional structure to align accountability and control.

When Steve Jobs returned to Apple, in 1997, Apple had this conventional structure (Podolny and Hansen 2020). In accordance with the multi-divisional doctrine, the organisation was divided into business units, each with its own P&L responsibilities. Believing that conventional management had stifled innovation, Jobs, in his first year returning as CEO, had fired all general managers of the business units, put the entire company under one P&L, and combined the disparate functional departments of the business units into one functional organisation that aligns expertise

Table 6.4 Apple's growth after adopting a functional structure

	Employees	Revenues
1997	8000	$7 billion
2019	137,000	$260 billion

with decision rights. This structure of experts-leading-experts rather than empowering general managers has become a characteristic of Apple's corporate culture.

Apple achieved an impressive growth in revenues during the two decades after Jobs' daring intervention of adopting a functional structure (Table 6.4).

This was essentially made possible by relentlessly focussing on design quality of the products created by Apple.

Jobs felt that those with the most expertise and experience in a domain should have decision rights for that domain and not general managers who lack the in-depth knowledge required. It is easier to teach an expert to manage than to train a manager to become an expert.

Apple's concept of experts-leading-experts encompasses three key leadership characteristics: deep expertise, immersion in details and willingness to collaboratively debate. To ensure that these attributes are ingrained into the corporate culture, Apple University was founded in Cupertino, California.

The history of Apple shows that, at least for License Giver organisations, the Multi-divisional Doctrine has definitely outlived its usefulness.

Implementation of the experts-leading-experts model is not at all easy. Those who dislike the Adhocracy structure will even feel less at ease in the experts-leading-experts model that brings along even more ambiguity. They should remain a License Taker that does not need to innovate. As the managing director of a French License Taker once confided: 'Je travaille pour vivre en France'. I work to live in France.

6.3   Transition from License Giver Back to License Taker Due to Continuous Decline of Innovating Capability

License Giver companies can lose their innovating capability for all kinds of reasons. The most common one, for sure, is what we may call the Succession Paradox (van Gunsteren 1995, p. 827):

If a great leader accepts a candidate for his succession, you can be sure the candidate is the wrong one, but when the right one is found the leader does not accept him.

Case: Pan Am, USA

Pan Am was founded in 1927 by Juan Tripp, a veteran pilot of the First World War and a businessman already involved in the airline industry. He was leading the company as CEO until retiring in 1968, leaving a solid company with assets totalling over $1 billion (van Gunsteren and Kwik, pp. 87–95).

After Tripp's retirement, Pan Am gradually declined until its bankruptcy in 1973. The rise and fall of Pan Am is well documented elsewhere and only mentioned here to point at the pattern: Four decades of inspiring leadership and growth followed by about two decades of decline and ultimately disappearing altogether from the scene.

Case: Lips Propeller Works, Drunen, The Netherlands

Under the leadership of Max Lips, the firm had grown from almost bankruptcy in 1934 to a sizable multinational corporation and a worldwide leadership in Ship Propulsion in 1974. See Figs. 6.1 and 6.2 taken from the book Lips Drunen published on the occasion of Max Lips' 40 years directorship.

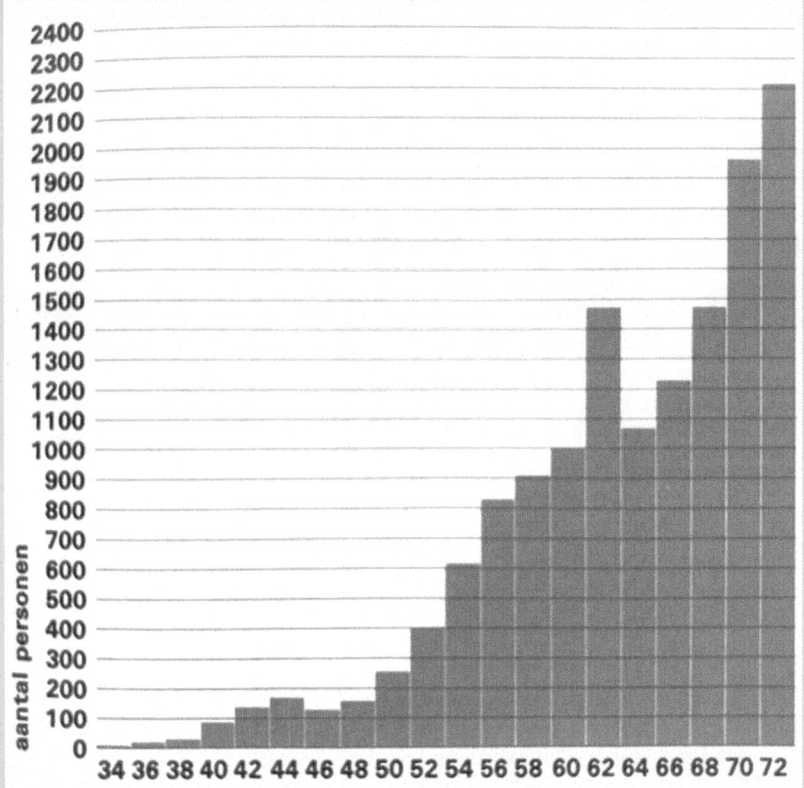

Fig. 6.1 Lips' number of employees in the period 1934–1973

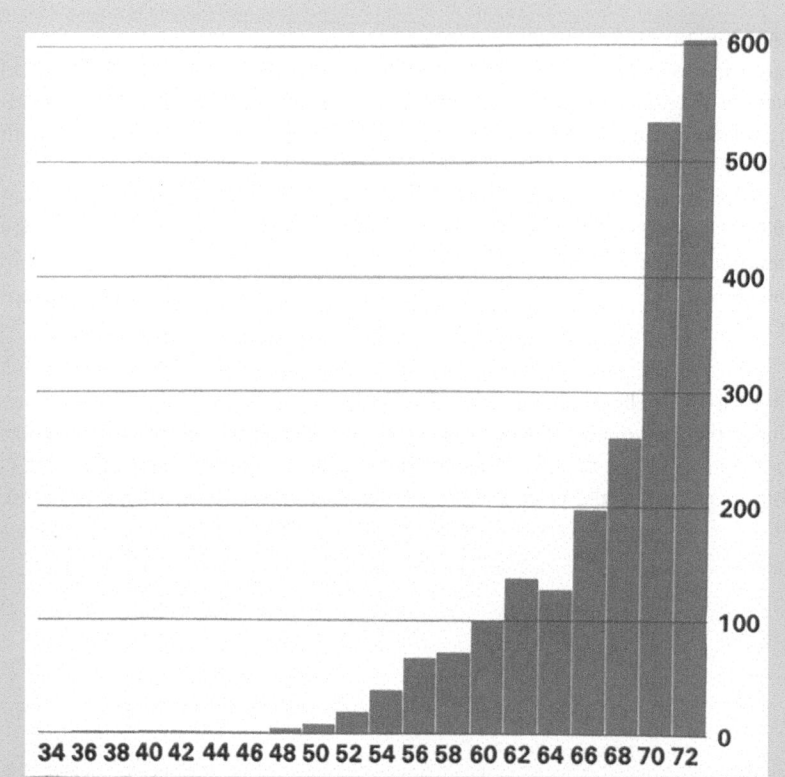

Fig. 6.2 Lips' revenues in the period 1934–1973. 1960 = 100

Max Lips' ideal of becoming a worldwide leader, a true License Giver, in ship propulsion had been achieved at last. But 40 years of spending an incredible amount of energy in his enterprise had taken its toll. When Max Lips turned 60 in 1973, he announced his intention to retire. His active role in managing the company came definitely to an end when in September 1975 an accident chained him forever to a wheelchair.

The continuous decline that followed is characterised by frequent changes in ownership and in composition of the executive and supervisory boards (Bult and Straatman 2011, pp. 143–147). Such organisational climate of continuous crisis is detrimental for innovation. From the achieved but still fragile License Giver identity, the company was forced back to a License Taker position and finally disappeared completely from the scene when the Finnish corporation Wärtsilä acquired full ownership in 2002. The name Lips disappeared, and the production was transferred in 2010 to China and Spain. All that is left in Drunen is a bunch of empty manufacturing halls.

The License Giver Business Concept in the Evolution of Industries

7

Industries emerge, develop and mature. In this chapter, we sketch in broad lines how the License Giver business concept fits into the evolution of industries.

Table 7.1 presents a summary of the evolution of an industry by distinguishing four different stages of development and their features.

The initial formation of industries was a result of the first industrial revolution. Cement and investment in the building of factories, costly steam engines to power them, and gas lightning to enhance capital productivity by enabling longer working hours. Steam engines powered milling and other machines that enabled the production of standardised parts. Industry structures remained vertically aligned with manufacturers inventing new products, but design and manufacture of all major parts remained in-house and custom to its own products and operations.

Maturing of core technologies like steam and fossil fuel engines, and mastery of the manufacture and design-in of standardised parts across different designs laid the groundworks for the eventual horizontal realignment of industries around specialist providers. A horizontally structured industry requires more complicated R&D skills and processes. It necessitates cross-company collaborative R&D coordination, design rules and processes, and IP licensing.

The process of industry restructuring over time is now explained for the ship, aircraft, vehicle and computer industries.

7.1 Ships

The *birth* of the industry of steel ships took place at the end of the nineteenth century. Until then, ships were built out of wood. In this first stage, the new material was incorporated into the empirical knowledge base of wooden ship design.

Table 7.1 Evolution of an industry from its birth to maturity

Stage	Feature
Birth of the industry	Creation of an artefact
Vertical alignment	Continuous improvement of the quality, the fitness for purpose, of the artefact
Horizontal alignment	Reaping the benefits of economy-of-scale and specialisation
Maturity	Artefact's adoption of new technologies

In the stage of *vertical alignment*, shipyards became the driving forces for continuous improvement of the quality of the artefact. The necessary supporting workshops were located on the premises around the slipways and docks: a machine shop for steel work, a carpenter shop for wood applications, a foundry for casting work, a draughting department for design work and so on. The process of converting materials into ships was kept almost entirely in-house. Subcontracting was close to nihil. A large workforce was necessary to keep delivery times within acceptable limits. For instance, the yard Harland & Wolff that built the Titanic in 1912 employed a workforce of almost 30,000 people.

The transition to the stage of *horizontal alignment* took off shortly after the end of the Second World War. Outsourcing and subcontracting became a popular practice. Steam turbines and diesel engines were no longer manufactured in-house but directly bought from specialised suppliers. Yards closed their foundries for casting propellers and anchors, terminated their production of diesel engines and outsourced many other items.

The driving force of these developments was the fact that specialised suppliers can reap the benefits of economy-of-scale and specialisation much better than their counter parts at the yards. Ultimately, almost everything was outsourced: scaffolding, painting, interior decorating and many more functions. Two main functions remained for the yards. First, integrating all these outsourced activities into a well-balanced *design* with a high degree of fitness for purpose. Second, *assembling* the parts into a ship that can be delivered to the end user. When there is nothing left to outsource the industry has entered the *maturity* stage.

7.2   Aircraft

The *birth* of the aircraft industry dates from the beginning of the twentieth century when the Wright brothers were the first to get an airplane in the air.

In the stage of *vertical alignment* that followed, aircraft companies like Fokker, Focke-Wulf and Messerschmitt, improved consistently airplane performance: speed, range, load capacity, comfort of the pilot and many more features. Outsourcing was almost nihil. All functions were executed in-house.

The transition to the stage of *horizontal alignment* started shortly after the Second World War. Like in shipbuilding, the engine was the first item to be outsourced to specialised suppliers like Pratt & Whitney, General Electric and Rolls

Royce. The leading aircraft manufacturers Boeing, Airbus, McDonnell Douglas outsourced consistently every item for which a competent supplier could be found. Nowadays, the leading aircraft builders only take care themselves of design and assembling, while outsourcing all other functions.

Aircraft builders that ignored the change of the rules for business success in the stage of horizontal alignment have disappeared from the scene.

Case: Fokker Aircraft

The history of Fokker Aircraft is impressive. Its contribution to the development of aero plane technology is unequaled. Their latest designs, the Fokker 27, the Fokker 50 and Fokker 100 were widely appreciated in the market. How could it happen that Fokker was nevertheless forced to terminate its business as an independent aircraft manufacturer? The main cause has been that Fokker continued to operate according to the rules of the era of vertical alignment, while the whole industry had moved into the era of horizontal alignment. The conversation the authors had with a senior Fokker executive illustrates the point. We asked him why Fokker did not establish License Takers in Brazil and Indonesia. Between Rio de Janeiro and Sao Paulo, flights took place every hour. The Fokker 100 would fit perfectly in that market segment and would not require an extension of the landing strip, as would be necessary for the Boeing 737. The Fokker100 would also fit well in the market segment of connecting the isles of Indonesia. A License Taker factory in Jakarta would be very welcome. The Fokker person then became emotional and responded that Licensing to developing countries was out of the question because the safety of the planes produced in License would be seriously affected. This could be a valid consideration in the vertically structured industry of the past, but not in the horizontally structured industry of the future. Fokker's management ignored that the rules for successfully conducting business had changed from outsourcing as little as possible to outsourcing as much as possible to whoever is the best in the field concerned, the License Giver with his License Takers. The safety of the planes would not be an issue provided the License Giver keeps properly control over quality assurance and control. Excluding licensing from the firm's strategic repertoire implied that Fokker was actually managing the firm with the mindset of the past era of vertical alignment, instead of embracing the rules of the new era of horizontal alignment. This made the demise of Fokker inevitable.

7.3 Vehicles

The *birth* of the automotive industry began in the1860s with hundreds of manufactories that pioneered with the 'horseless carriage'.

In the stage of *vertical alignment*, the performance of cars was steadily improved. Competition was initially focused on design and later on mass production. The transition to the next stage began with the outsourcing of tires and engines. After the Second World War, outsourcing really took off, indicating that the stage of *horizontal alignment* had been reached.

Most components and subsystems are nowadays manufactured by Original Equipment Manufacturers (OEMs). The OEMs serve the whole car industry. For instance, the OEM of mirrors delivers mirrors to all car makers, the OEM of steering wheels delivers steering wheels to all car makers and so on. Cross licensing has become normal in the industry and is often enforced by car makers who wish to maintain the balance of power among the OEMs that serve them.

7.4   Computers

The birth of the computer industry happened around WWII with the invention of the transistor. Till the 1980s the industry was vertically aligned with companies like IBM, Sperry Univac, NCR and Olivetti.

The stage of *vertical alignment* included the following elements: chips, computer, operating system, application software, sales and distribution (Grove 1997, p. 40) (see Fig. 7.1).

The combination of a company's own chips, own computers, own operating systems and own application software were sold as a fully integrated solution by the company's own sales people.

Then, with the launch and rise of the personal computer (pc), the microprocessor and pc operating system emerged as the defining building blocks of the industry. Economy-of-scale and specialisation made manufacturing computers cost-effective.

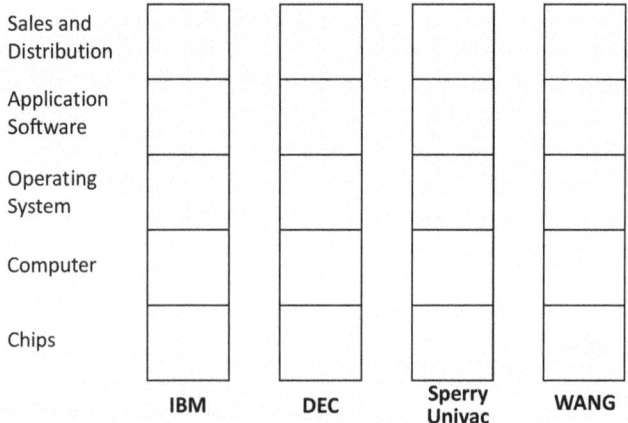

The Vertical Computer Industry - Circa 1980

Fig. 7.1 The Old Vertical Computer Industry-Circa 1980

Under the new horizontal industry structure a consumer could pick from a wide range of computer system configurations from different pc manufacturers built around unifying component standards, especially the microprocessor and operating system building blocks. These were also available through an expanding number of independent channel options, where the consumer could also grab and install their choice applications from the shelf. A computer system produced the old way would cost ten times more.

Newly emerging networking solutions emerged alongside, enabling the personal computers to connect across local and wide area networks powered by small cost-effective servers (instead of large mainframes). This 'open' client-server system architecture started to quickly replace the old 'closed' mainframe-centric architecture.

Over time, this changed the entire computer industry. A new horizontal industry had emerged. See Fig. 7.2 (Grove 1997, p. 42).

Companies that had done well in the vertical computer industry found it difficult to adept. To others, the new order provided an opportunity to raise to prominence. Compaq, for instance, became the fastest Fortune 500 company to reach $1 billion in revenue, and soon thereafter Compaq was followed by Dell to then become the fastest to $1 billion.

The continuing transition from vertical to horizontal alignment generates an ever-improving cost/quality ratio in the computer industry. Figure 7.3 presents an example from the computer industry (Grove 1997, p. 67).

Other industries show similar trends. Bicycles, household appliances, CV's, TV's, cars, trains, printers, etc. all continue to become better and cheaper as a result of the fundamental rule in technology that says that *whatever can be done will be done*. As a corollary, a design leadership will evaporate over time when nothing is

The Horizontal Computer Industry – Circa 1995

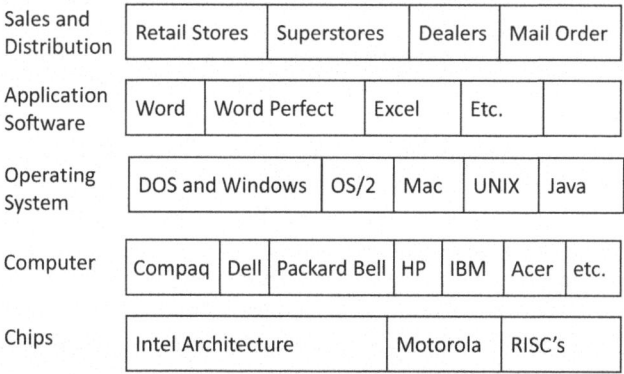

Fig. 7.2 The New Horizontal Computer Industry-Circa 1995

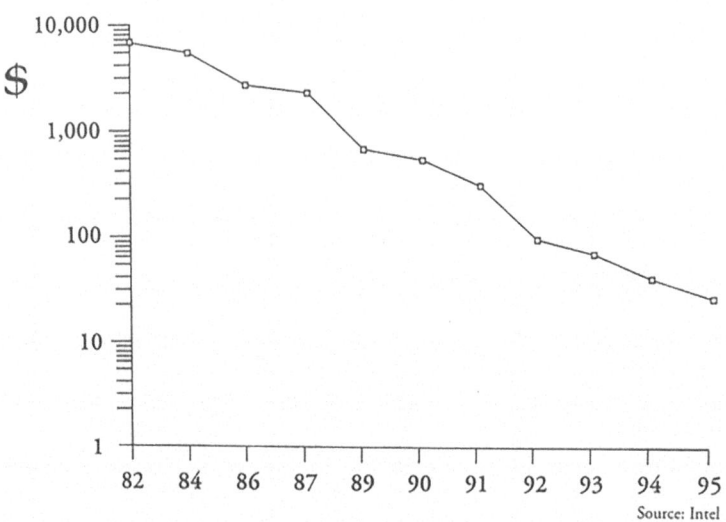

Fig. 7.3 The ever-improving cost/quality ratio in the computer industry

done with it. Hence our recommendation to always consider licensing to be a serious strategic option.

The relevance of the License Giver business concept is that it facilitates companies to change their strategic viewpoint on licensing as required by the increasing *horizontal alignment* of their industry. The more License Givers together with their License Takers become available, the more outsourcing becomes possible and the more the benefits of economy-of-scale and specialisation can be reaped.

During the process of an industry moving from a vertical to a horizontal alignment the original leaders need to demonstrate agility to adjust their approach to stay successful. In principle three options are available to them. The first option is to specialise in overall system integration, and refocus on overall system design, quality control and efficiency. Probably the best-known examples can be found in the automotive industry where a number of old, previously vertically integrated, car brands still stand tall and are highly successful.

The second option is to maintain design leadership in one or more of the horizontal layers. This was an option available to IBM early on, but IBM entirely misjudged the rise to prominence of the personal computer and the 'client-server' system architecture.

The third option is to try to maintain their vertically integrated system, but neither of the authors can think of any successful example, whilst there are many examples of admirable old-order leaders that resisted the inevitable restructuring, only to succumb and disappear from the scene.

7.5   Industry Platforms

In the computer industry, a further new industry and innovation dynamic emerged: the phenomenon of industry platforms. For some companies it offered the opportunity to establish themselves as platform leaders.

Let us consider the hypothetical case of three License Givers and their associated License Takers whose products could be the cornerstones of a new product with high market potential. The implementation of that opportunity would require collaboration of the three License Giver organisations that are complementary to each other. For the coordination of their efforts, they form an alliance which is called an *Industry Platform.*

Industry Platforms constitute a business model for reaping the benefits of *complementary* features of their participants (Gawer and Cusumano 2013, pp 417–433).

Performance of Platform companies has been impressive. The opening statement of the article 'The Future of Platforms' (Cusumano et al. 2020) reads

> *The world's most valuable public companies and its first trillion-dollar business are built on digital platforms that bring together two or more market actors and grow through network effects. The top-ranked companies by market capitalisation are Apple, Microsoft, Alphabet (Google's parent company), and Amazon, Facebook, Alibaba, and Tencent are not far behind. As of January 2020, these seven companies represented more than $ 6.3 trillion in market value and all of them are platform businesses.*

The authors of the article report furthermore:

> *Most platforms lose money, but platforms that dominate their markets have been extraordinary successful. When we compared the largest 43 publicly listed digital platform companies from 1995 to 2015 with a control sample of 100 non-platform companies in the same set of businesses, we found that the two samples had roughly the same level of annual revenues (about $ 4.5 billion). But platform companies achieved their sales with half the number of employees, were twice as profitable, were growing twice as fast, and were more than twice as valuable as their conventional counterparts.*

In the process of examining the 43 success stories, the authors identified 209 platform companies that were their direct competitors but failed or disappeared as independent companies. Some successful platform companies maintain their powerful positions for decades, but the creation of a platform by itself is no guarantee of long-term success. Timing is crucial to attain the leadership in an industry platform that is required for lasting success.

Horizontal alignment enables products to become better and cheaper, but does not generate a variety of innovation and new products with high market potential as industry platforms enable. The key difference is that within an industry platform the end use of the end-product(s) is not fully determined. Instead, the 'openness' of key platform ingredient technologies and their interfaces allow others to innovate around it, allowing for the creation of a greater variety of final end user products with unique qualities.

Intel and Microsoft were two defining platform leadership companies for the computer industry for about three decades. The microprocessor and operating system were critical components for the personal computers, and both developed as open standards with interfaces for others to develop compatible hardware and software components around it. As the personal computer and server markets exploded, both companies developed new technical and commercial programs and organisational capabilities that enabled them to lead the industry across many generations of their microprocessor and operating system architectures. While Intel and Microsoft also competed in certain areas of the platform architecture, in the end they always managed to prevent fall-outs that could have threatened their combined platform hegemony, often referred to as the 'Wintel' platform.

Over the last decade, though, the emergence of the smartphone proved to be a disruptive client form factor for the Wintel platform. In the early days the first smartphones did not require near the compute power that had come to define the desktop and notebook computers of the days. This presented a major dilemma for both Intel and Microsoft. The 'weight' of their most advanced legacy processor and operating system architectures made it cumbersome to design phones with them, especially in light of the size/weight/cost constraints the market put on mobile phone designs. In addition, the phone industry had also just witnessed how Intel and Microsoft had come to lead the computer industry and with it pull in a lion share of the profits as well, something they were not immediately keen to see happen in the phone industry.

Fast forward to today and neither Intel nor Microsoft established a foothold in the smartphone business. Which is mostly dominated by the Android operating system and ARM microprocessor architecture. Both these alternatives presented the industry with platforms with attractive licensing approaches at a higher degree of 'openness'. Android and ARM both allow License Takers to make certain changes to the core architecture, more so than Intel and Microsoft did. In the early days of the personal computer it can be argued that such a level of openness and flexibility would have led to breakdown of interoperability, and thus fragmentation of the industry and insufficient economy-of-scale to fuel the growth that it realised. Three decades later, though, with the accumulated industry experience to design IT client devices around platform standards, the mobile phone industry was able to handle the increased platform openness without losing the platform stability and interoperability.

Intel and Microsoft pioneered Industry Platform leadership, still dominate the large and profitable pc and server business. Both Intel and Microsoft recognised the potential of the smartphone and spared no expense to enter. But neither got the designs and timing right, nor considered to adapt their License Giving approach, to gain a sufficient foothold in the phone industry to build from. They were pioneers and showcase examples for assuming and maintaining an industry platform leadership position, but also testament to the fact that early success is no assurance for future success, not even in emerging adjacent industries.

Governmental Interference

8

8.1 Useful Interference

Indirect governmental support to innovation, in particular regarding legislation and infrastructure around clusters of competence, is useful and should be welcomed.

Mega infrastructure projects with innovating content like the Dutch Delta Works after the flood in 1953 in which 1836 people lost their lives, need governmental guarantees to be financed.

Case: Airport Island in the North Sea
Amsterdam Schiphol Airport is faced with the dilemma:

1. To remain a hub, an airport for intercontinental flights, the number of flight movements (landings and take-offs) per year has to be above a threshold in the order of 600 thousand flight movements.
2. To limit environmental effects, noise and air pollution, the number of flight movements per year has to be kept below a threshold in the order of 400 thousand flight movements.

   Both thresholds are clearly incompatible, causing an endless debate on the compromise that has to be reached.

   The dilemma could be resolved once and for all by an airport island in the North Sea (Binnekamp 2005, pp. 91–96). The return of investment is quite good (van Gunsteren 2003b) and the concept of an artificial airport island in the sea has already successfully been applied in Hong Kong and Japan. The investment of about 40 billion Euros has to be guaranteed by the government. So far, political support has been lacking. Therefore, before anything else, a

---

© The Author(s), under exclusive license to Springer Nature Switzerland AG 2022
L. A. van Gunsteren, A. G. Vlas, *The License Giver Business Concept of Technological Innovation*, Future of Business and Finance,
https://doi.org/10.1007/978-3-030-91123-2_8

Fig. 8.1 To acquire public funding, the issue has first to be moved from the IMPORTANT to the URGENT

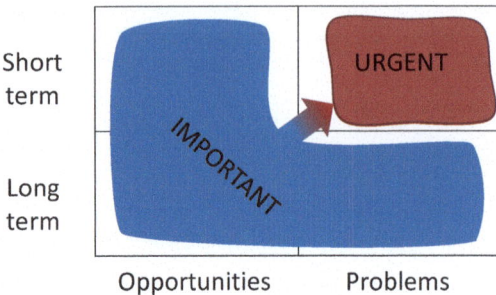

Short term

Long term

Opportunities   Problems

constructive political debate is needed. The problem is to get the issue on the political agenda, not the technical or economic feasibility of the proposition.

In the country's current fragmented political constellation, discussions tend to be about the division of public revenues rather than how to generate them. Large innovative public-funded projects have first to be moved from the IMPORTANT to the URGENT to become feasible (Fig. 8.1).

Insiders were well aware that the dikes were not high enough to withstand the simultaneous occurrence of a spring tide and a heavy storm. But the urgency to act needed the disaster of the flood in 1953 to become serious.

In a democratic society, public debate is the only way to resolve such problems of public affairs (van Gunsteren, H.R., 1976, p. 154).

8.2 Counter-Productive Interference

Direct governmental interference in existing industries is always counter-productive in the long run. Direct subsidies to sunset industries provide only short-term relief. On the long run, their distortion of competitive market conditions weakens the resilience of the subsidised corporations. General Motors, automobiles, in the USA and RSV, shipbuilding, in the Netherlands are notorious examples.

Politicians display a strong belief in the erroneous *rational view of innovation*. They cannot accept the reality that their power is extremely limited in regard to how, where and when innovation takes place.

Case: Closure of the Shipyard IHC Gusto, Schiedam, in the Late Seventies
As part of his plan for restructuring the country's shipbuilding industry, the minister of Economic Affairs, Ruud Lubbers, wanted a merger of all Dutch offshore companies into one national offshore corporation. The new company would be part of Rijn-Schelde-Verolme (RSV) and get at its disposal the most modern yard of RSV, Verolme Rozenburg, as well as IHC Gusto's superior

engineering capability. Gusto's managing director was invited to become the CEO of the new enterprise. The Gusto yard would be closed. Lubbers was prepared to pay the owner of the yard, IHC Holland, Hfl. 120 million tax-payer's money to cover the cost of the closure. Dick Smit, president of IHC, considered this to be an offer he could not refuse, so the deal was made.

Lubbers and Smit underestimated the sanction power of quitting that employees can always mobilise against the formal power of their bosses. The reaction on Lubbers' plan was an exodus of key personnel, including the managing director of the yard. His view was:

1. The closure of the yard implied a betrayal of the employees who would unnecessarily lose their job. The investment in the fixed assets of the yard of 50 million Dutch guilder, already approved by the Supervisory Board, would be enough to get the yard's cost price at a competitive level.
2. The 30% reduction of shipyard capacity in the country as aimed at by Lubbers would only reduce the Dutch world market share from an insignificant 2% to 1.4%.
3. Offshore is an international business dominated by the Americans and the French. The yard maintained close relations with the latter. The image of being a national company, dependent on support from the government, could become a handicap for getting orders from the global offshore market.
4. The trend in offshore equipment was towards installing manifolds at the sea bottom instead of operating from a vessel. The fixed cost of the depreciation of the Verolme facilities for large ships would be a burden rather than an asset.
5. The post-merger culture clash with the bureaucratised RSV would be prohibitive for successful innovation for at least a decade.

Although officially downplayed, the exodus of key personnel raised serious doubts about the viability of the offshore merger and was reason to involve former minister of Economic Affairs, Harry Langman and McKinsey who soon concluded that the merger plan had to be abandoned.

Similar to Shockley's 'Traitorous Eight' from Chap. 3, seven departing engineers ended up forming their own company together.

They founded the consultancy firm Marine Structure Consultants (MSC) as a working company of the Royal Boskalis Westminster Group whose CEO welcomed them with the words: 'You can't lose money on brilliant engineers'. MSC became part of IHC Caland in 1988. What remained of the engineering capacity of the yard also survived as an engineering consultancy firm operating under the name Gusto Engineering. The two companies started to collaborate in 2003 and merged in 2011 under the name GustoMSC, which still is a leading company in offshore engineering.

This history shows that Gusto's excellence in engineering made it possible to become a License Giver type company in its own right like Sulzer. Gusto had already granted licences to overseas yards for building drill ships of their design. Instead of nurturing this opportunity, Lubbers spent Hfl. 120 million of public funding to get the yard closed. This blunder never became an obstacle for becoming Prime Minister of the Netherlands.

Case: The Innovation Platform in the Netherlands (Huibregtsen 2017)
In 2003 the Dutch cabinet installed an Innovation Platform under the chairmanship of the Prime Minister Jan Peter Balkenende, including some 20 prominent representatives of business, education, government and science. The platform was created to address the so-called knowledge paradox. Although the Netherlands scored relatively high on scientific research in international comparisons, its share of innovative products and services was low.

The task of the platform was to stimulate innovation and entrepreneurship in the Netherlands. Its target was to reach a global top five position in higher education, research and innovation.

The efforts of the platform resulted in a reduction of barriers for foreign knowledge workers to enter the Netherlands. The introduction of innovation subsidies for small and medium sized companies, and the creation of an agenda for targeted investment in knowledge. To open opportunities for further expansion key economic areas in which the Netherlands already occupied a strong position were designated.

In 2007 a new platform was instituted by the next cabinet which was ultimately disbanded in 2010. When it became clear that also the new platform failed to generate the projected increase of innovations, it was disbanded.

Case: Submarines for Taiwan
The 1991/1992 Dutch debate on the sale of submarines to Taiwan has been described by Ingrid d'Hooghe (1992). This debate, preceded by two earlier debates on the same issue in 1980 and 1983, concerned the export permit for the sale to Taiwan of four submarines (with option of six more) to be built by the Rotterdamse Droogdok Maatschappij (RDM). The total value of the order was estimated at Hfl 2.8 billion (1.3 billion Euros).

In d'Hooghe's article the role of the Minister of Economic Affairs, Frans Andriessen, is described as follows: 'The position of the Ministry of Economic Affairs was ambiguous. In 1984, at the restoration of full diplomatic relations with China after the second debate in 1983, the Netherlands had tailored its policy to the expectation that trade with China would rapidly increase as

Fig. 8.2 RDM expertise in action

China was opening up to the West. These hopes proved to be false, and so it was likely that the Ministry of Economic Affairs would give serious consideration to strengthening relations with Taiwan. However, Minister Andriessen eventually spoke out against the sale'.

Apparently, the minister did not realise that for the RDM this meant the end of being able to maintain itself as a License Giver (Fig. 8.2).

A decade earlier when the RSV-crisis occurred, the RDM's expertise in welding and finite elements-based stress analysis of high carbon steel materials as applied in pressure vessels of nuclear installations and submarines, was well recognised. Designing and building submarines were declared to be RDM's core business. The track record of the two submarines that were delivered to Taiwan in the eighties and are still operational, the performance of the submarines of the Royal Netherlands Navy itself, and the history of the Dutch submarines in WWII, would highly contribute to the credibility of any sales proposal.

The License Giver's top priority is to get as many as possible applications of its design executed, no matter the circumstances like ownership, political situation or financial return. If the License Giver does not succeed in this respect, recovery of RD&D costs will not be possible in the long run. This means that RDM simply could not afford to miss this order.

Another valuable License Giver was lost by the incapability of a Minister of Economic Affairs.

Case: Royal IHC

In April 2020, the Dutch government announced to grant a subsidy of Euro 400 million to keep Royal IHC afloat. Minister of Economic Affairs and Climate, Eric Wiebes, and State Secretary for Finance, Hans Vijlbrief, wrote in their letter to the House of Representatives of April 30, that they felt they needed to intervene. The maritime journal SWZ summarises the letter as follows: '*IHC plays a strategic and innovative role in the maritime sector and failure to do so would undermine the robustness of the entire sector and have a major impact on the international competitive position of the maritime manufacturing industry. Secondly, a lot of jobs would be lost at a time when the government is doing everything in its power to keep the economy going and to minimise job losses.*' This impressive rhetoric concealed what was actually happening: wasting hundreds of millions of Euros public funding for knowhow that is readily available and maintaining jobs for which there is no need. Building trailing hopper and cutter suction dredgers is relatively simple and does not involve much advanced technology as needed for building submarines or drill ships.

To summarise, politicians and civil servants should stop:

- Telling inventors what they have to invent.
- Giving-in to blackmail as to which countries firms are allowed to conduct business with.
- Subsidising sunset industries.
- Subsidising jobs that are not needed.

8.3   Relevance of the License Giver Business Concept for China's Industrial Policies

China is not only a nation state but also a civilisation with fundamental differences compared to current Western-style democracies. China has a 4000-year history rich with cultural and scientific contributions, and can be credited with an impressive list of major inventions. After WWII today's democracies decisively moved away from autocratic governmental models. Propelled by an accelerating train of technical developments they prospered, and saw their economies and companies globalise, supported by international institutions like the United Nations, the World Bank, the WTO. Following a few humbling experiences with invasive actions by Japan and a few Western nations, China instead returned to its more seclusionist approach to international relations, and maintained a firm autocratic model under the rule of the Chinese Communist Party (CCP).

Many interpreted Deng Xiaoping's re-opening of China, and later its entry into the WTO, as China embracing not only our products and technologies but also our values of economic and individual freedom, and even democratic values. This was not an improbable scenario when looking at the historic developments in nearby Japan, Korea and Taiwan. Instead, the CCP has consistently maintained a central and pervasive presence across its institutions, society and daily life. The CCP did so with the objective of 'providing China and its people *security* and *stability* in a turbulent world'. Other possible objectives like growth, fighting air pollution and climate change, individual freedoms, etc., play a subservient role. As for the legitimacy of the CCP's rule, a July 2020 poll from the Ash Centre at Harvard's Kennedy School of Government indicated a 95% satisfaction rate among Chinese citizens with the Beijing government (Mitter and Johnson 2021). Importantly, and in contrast to what many outsiders believe, the Chinese people seem of the opinion that the massive poverty reduction, infrastructure investment and technical progress were made possible because of, not in spite of, China's authoritarian model (Mitter and Johnson 2021).

China is expected to surpass the USA as the world's largest economy. This requires the economy to cross the treacherous 'middle-income gap', i.e. continued GDP growth in spite of rising labour costs, and a structural shift towards increased domestic consumption away from over-dependence on exports. China's homogenous population of 1.3 billion people, and increasingly affluent and highly educated population, represents a huge home market offering lots of opportunities for new products and services and entrepreneurial initiatives. The pace at which the development of China towards becoming the world's number one economy will take place depends to a large extent on where and how it pursues independence of foreign technologies. In this respect, the License Giver business concept could become relevant.

Technology, innovation and Intellectual Property (IP) play an important role in China's contemporary central industrial policies. The ambition is to develop from being the manufacturer of the world towards also becoming the research laboratory of the world. The manifesto MIC2025 (Made In China, 2025), backed by an estimated $500 billion in government funds (Black and Morrison 2021), specifies targets over three decades as to how certain industries should develop and where acquisition of foreign technology remains necessary (Overdiek and Arregui Coka 2020). Home-grown technologies are rightly seen as a key ingredient to overcoming the middle-income gap, and maintain growth and *stability*. A stated goal and parallel driver for the CCP is to gain technical control and independence in support of the overarching *security* objective. The ascent of China to the higher rungs of the technology ladder has become a very controversial topic in the media and global international relationships. In this highly complex geopolitical and economic matter the License Giver business concept can be relevant.

The USA still enjoys technological superiority, although that superiority may be dwindling faster and more irreversibly than one would expect. Patents play a central role, as an economic system to manage Intellectual Property rights, and as a measure of technical prowess and trends. China has bolstered its patent regime, to the benefit of both foreign and domestic companies. It is also actively pushing to increase patent numbers (in its '13th Five-Year Plan for Economic and Social Development' it aimed for 12 invention patents owned per every 10,000 people in China), resulting in a stunning 3.5 million applications in 2016 (Prud'homme and Zedtwitz 2018). At a next level of granularity, though, the general sense is that many of these patent filings are still relatively weak in terms of technical strength. Strong inventions that represent fundamental technical breakthroughs require the right environment to allow creativity and inductive thinking. In this respect the conditions in China are unfavourable.

Driven by fast-moving Chinese entrepreneurial spirit and positive experience of numerous economic and societal changes over just 30 years, China is embracing new technologies at a pace much faster than the USA and Europe (Zak Dychtwald 2021). As a result, China has excelled at commercialising certain innovations, by experimenting with and learning about new usage models, and developing accompanying business models. E-commerce solutions and its widespread adoption are prime examples. There is lots to learn for European and US companies by participating in the China economy.

The integration of quality and innovation into the Chinese civilisation has been extensively treated in the book 'China's Need for Small Northern European Friends' (van Gunsteren, F.F., 2011). The author identifies, and illustrates with personal experiences, a number of historically grown cultural impediments towards creating an environment that is conducive to innovation. The article 'Why China cannot innovate' (Abrami et al. 2014) comes to a similar conclusion, especially pointing to the negative impact of increased levels of presence and directional influence of the CCP in companies as well as educational institutions.

The following five counter-productive governmental interferences have been identified by Anil K. Gupta and Haiyan Wang in their article 'How China's Government Helps—and Hinders—Innovation' (HBR, November 16, 2016):

1. The bulk of China's government R&D funds are allocated on the basis of political connections rather than merit as judged by independent scientific panels.

2. China spends relatively little on basic research compared with the OECD-economies (4% vs. 17% of total).
3. Priority on the quantity over quality of patents.
4. The 'Great Firewall of China' makes it difficult for Chinese researchers to access global information. Chinese researchers cannot access Google Scholar.
5. Foreign companies feel pressured to transfer technology to gain market access and are at a disadvantage regarding IP-related judgements in Chinese courts. The outcome has been that, while almost all western technology giants have R&D labs in China, bulk of what they do is local adaptation rather than developing next generation technologies and products.

The current trade conflict between China and USA constitutes a further road-block to creating circumstances in China that allow emergence of innovation eco-systems such as Silicon Valley.

The USA is still the undisputed 'patent superpower' with most patents in 50 of 58 cutting-edge technologies, and continues to hold this position in the medium term, but China is catching up quickly. China ranked among the three countries with the most world class patents in 42 of the same 58 technologies examined, and holds most in five technologies in the fields of nutrition and environment. (Bertelsmann Stiftung, 2020). Patents constitute only one indicator technical progress along with the numbers of Nobel Prizes, peer-reviewed publications, PhDs and citations.

China has highly successful manufacture ecosystem clusters, like the consumer electronics cluster in Guangdong Pearl River delta. Several have proven themselves as successful development centres and sources for incremental and market-driven innovations.

Governmental industrial policies are pursued by China and the USA to bolster competitiveness and innovative capability. The current regimes, though, both increasingly resort to measures that are focused at tightening control by the government. It seems that China's central and the US's federal governments are unaware of the innovation paradox: Innovation cannot be directly managed, since creativity and inductive thinking, as opposed to deductive thinking, cannot be imposed top-down.

The government should consider focusing on creating circumstances that are conducive to innovation and entrepreneurship. For example, the government could designate specific areas having the potential of becoming clusters of competence in various trades: an automotive cluster, a textile cluster, a wine yard cluster, etc., all situated near a university and already being involved in the trade concerned. These clusters not only should be supported by appropriate infrastructure but also as be granted the psychological freedom and safety which are indispensable conditions for being creative. Sooner or later, some of the clusters will develop into License Giver companies with more fundamental breakthrough innovations and worldwide design leadership.

8.3.1   The 'License Giver and License Taker Structure'
as an Idealised Design

In regard to the relationship between China and USA, it may be worthwhile to think in terms of an *idealised design*, i.e. thinking backward from an ideal situation in the future towards the present, as opposed to thinking in terms of near term-steps to improve the current state-of-affairs. The ideal would obviously be that the Chinese civilisation and the Western democracies co-exist in peace while neither of the two tries to change the other. In this ideal situation, American License Giver companies would have their License Takers in various provinces of China and Chinese License Giver companies would have their License Takers in various states of the USA.

To make this possible, there needs to be an IP regime of patent legislation that recognises IP ownership.

During post-WWII early industrialisation of Japan, and recently China, more developed countries and their companies at times turn a blind eye to patent infringements by companies from those countries. In some cases the IP owner considered the technology in question as less advanced. Or there were other reasons, some naïve, to not pursue recognition of patent violations. As China is now catching up technologically, infringements increasingly concern more advanced technologies. Consistent adherence to patent legislation by both source and user is conducive to diffusion of innovation. Inconsistent use of rules renders them useless for the implementation of the License Giver business concept. Policies of Chinese companies to build patent fortresses to fortify a monopoly position and to allow, or even stimulate, infringement of patents for domestic use, are counter-productive in the long term.

Within a regime of patent legislation, License Takers can be foreign, domestic or joint ventures. The entity structure is best left up to market dynamics and companies to figure the most effective operational structure and resourcing model for the venture. In developing economies, governments often prescribed JV structures for foreign companies to participate in the market. In general, looking at the failure rate of JV's, this is a misguided attempt to increase participation by local players on an equity level. In addition, and detrimental to the proposed approach, governments have also been known to allow, or even encourage, inappropriate use of foreign IP by domestic JV partners.

There may be some specific technologies that are viewed as strategic or as a national security risk. For these a 'decoupled' technology strategy may be unavoidable. But there are many more products and technologies that are not strategic.

Adoption of a principle strategy aimed at broad technical decoupling of economies would represent a reversal of fundamentals that powered post-WWII prosperity and stability, globally and within China. A strengthened judicial IP regime and sharp increase in patents in China is a welcome development, especially when accompanied by a parallel increase of patent filings by these same companies in other major markets. Such international filings would represent a trend towards a more balanced global spread of patents, and an indication of intent to also use these patents in other markets. See Fig. 8.3 as an example of filing trends in the USA at

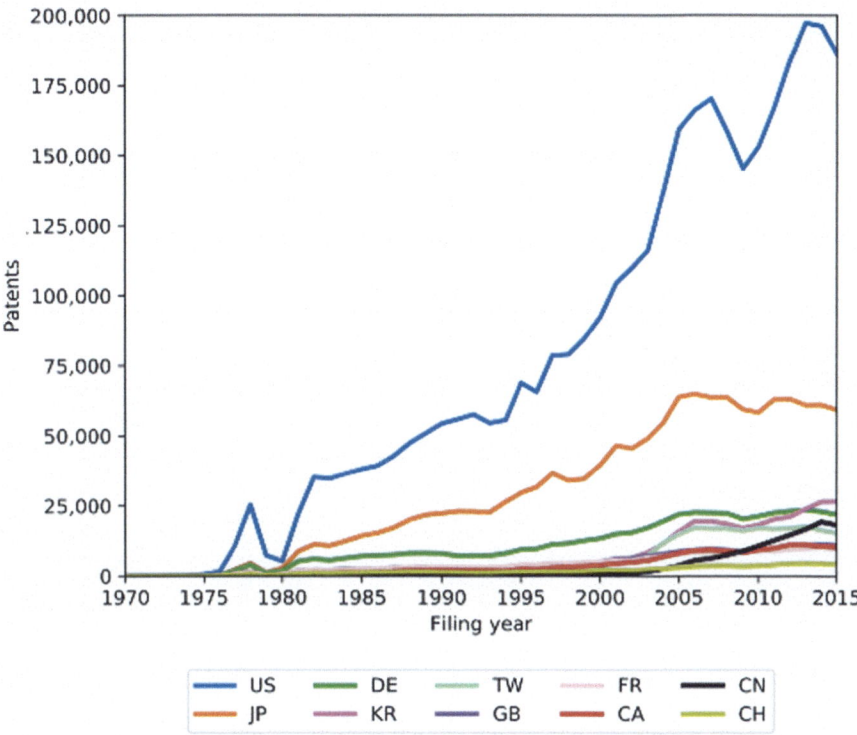

Fig. 8.3 The total number of patents filed at the USPTO each year by inventors in a given country (Webb et al. 2018)

the United States Patent and Trademark Office (USPTO) (Webb et al. 2018). If there is no parallel increase of patent filings in other markets, it can be interpreted as an indicator of a different intent. Namely a limited geographical use and thus as a tool towards technical decoupling by using patents to keep foreign competitors out of a market. This would be another example of a serious departure from the approach recommended here.

Patents define a judicial relationship between a licensor and licensee. The License Giver – License Taker relationship entails much more than a mere licensor-licensee agreement. The know-how that the License Giver transfers to his License Takers comprises more than can ever be specified in patent descriptions. The License Giver determines the quality assurance to which the quality control of its License Takers is subjected. Patents play only a limited complementary role in the License Giver business concept. Implementation of the License Giver concept requires respect for each other's roles and responsibilities, as becomes clear from articulating an "idealised design", a blue print of a desirable future state. This approach to international cooperation could possibly change the current hostile confrontation and zero-sum mindset communicated in recent policy statements by the USA and China. And ultimately reach a stable mode of collaboration based on mutual trust and respect. This point of view may be discarded as naïve but, unfortunately, there is no other way.

Excellence Through Competence

9

9.1 Tolerance for Failure Combined with Intolerance for Incompetence

Suppose you are commander of a submarine; would you allow incompetent crew members to come on board? Of course, not. You are well aware that any failure to perform can cause the whole ship to sink. For this reason, submarine recruitment procedures are extremely severe. The submarine unit in a navy is an elite corps.

The same holds for the innovating License Giver organisation. Every employee should have the competence required for the task entrusted to him or her. If such competence is lacking, the employee must leave or be transferred to a less demanding position. It is quite common, however, that an employee who displayed great loyalty in the past but failed to upgrade his expertise in due time is allowed to stay. The story of the Wing Nozzle is an illustrative example.

Innovation requires a tolerance for failure. The genuine daring attempt is what counts, not the result. But failure due to incompetence of the people involved should be considered to be unacceptable. A tolerance for failure requires intolerance for incompetence (Pisano 2019). Failures can be very useful to learn from. The lessons learned from the failure of the slotted nozzle have been crucial for the successful development of the Wing Nozzle. Without the preceding experience of the Slotted Nozzle, the Wing Nozzle would never have been invented.

The process of innovation is too fragile to allow incompetence of key decision makers. A strategy of becoming a License Giver organisation is bound to fail if its management is reluctant to remove employees that are lacking the required competence for their job.

How can one assess if someone is suited for a specific job? Procedures like management by objectives, yearly appraisals by bosses, peers and subordinates are useful but not enough.

L. A. van Gunsteren, A. G. Vlas, *The License Giver Business Concept of Technological Innovation*, Future of Business and Finance, https://doi.org/10.1007/978-3-030-91123-2_9

Information from these procedures has to be complemented by direct observation, ongoing candid coaching, and investing in people development. The following quote from van Gunsteren's book on leading professionals, clarifies the issue (van Gunsteren 2012):

> *When a subordinate asked me why his salary was lower than that of a colleague, I used to give a straight answer: 'Because he is better than you'. No diplomatic wording to soften the message. Of course, I had to be sure that my judgement was right. That means I had to be aware how difficult his job was and how well he was executing his tasks. To this end, I had to pay genuine attention to the work of subordinates without any fear of details.*
>
> *To be a good boss, it is more important to understand and recognise the value of the work of subordinates than to be a nice person.*

If the company culture is to be nice rather than candid, the excellence required for successful innovation will never be reached. Such company is advised to stay a Jobber or License Taker and refrain from any innovation ambition.

The requirement of competence holds for all employees, not only the academically educated engineers, as is illustrated by the following examples of Lips Propeller Works.

Example 1 The Blind Telephone Operator
The telephone operator was blind. He knew by hart the name, the telephone number and the function of every executive on the premises. This enabled him to connect incoming calls to the right person faster than anybody else.

Example 2 The Alert Guard
A guard on his round during the night hours saw someone he could not recognise as one of our regular programmers printing the listings of the highly confidential FORTRAN propeller design programs. He alerted the management. The caught spy turned out to be an American who had shortly before joined the sales department.

Innovating companies recruit consistently the best talent they can, regardless their envisioned rank and position in the organisation. Competent employees are the cornerstones for achieving the excellence as required for successful innovation.

9.2   Competence Maintained and Upgraded

Competence erodes over times as a result of technological progress. Riveting, once an essential skill in steel ship building, was replaced by welding technology. Drawings were replaced by plotters, which were later replaced by computerised storage of numerical data.

To remain relevant, competences must be maintained and upgraded. That means employees must be prepared to learn new things and acquire new skills when these become relevant in their job.

Example 1 Never Too Old to Learn: From Blue Collar Worker to Computer Programmer
A skilled factory worker had an accident that injured him severely. He was, therefore, transferred to the drawing office. When the drawing work was gradually automated by means of plotters and computers, a personnel stop was announced by the head of the department along with announcing company support in retraining for other work. The employee, who was in his early forties, reacted to this announcement by expressing a desire to become a computer programmer. To learn computer programming to the required level took him about three times longer compared with people in their twenties. But his subsequent performance in the computer department was more than satisfactory. Whenever there was a problem with computer programmes related to the drawing office and the factory, nobody else could match his performance.

Example 2 From Engineer to Manager
It is not uncommon that engineers become managers later in their career. In general, they do not have any difficulty to acquire the competences required in their new job. Engineers can become general managers. But the reverse, general managers becoming engineers, is almost impossible.

The Human Resources (HR) policies of the innovating organisation should be conducive to recruiting and retaining high-flyers, people who really make a difference.

Implications of this requirement are as follows:

- The employer offers work that contributes to the self-actualisation of the employee.
- The employer facilitates learning and acquiring new skills that have become relevant in the employee's job, such as learning the language of the country where an employee is located.

Effective learning requires a strong desire to learn. This means that the learning activities supported by the employer should not be standardised but tailored to individual needs.

Epilogue

The long-term impact of innovation on our prosperity and welfare justifies reflecting on the question what impedes or enhances the desirable creation and diffusion of innovation. For sure, there is no lack of ideas. More promising propositions are available than can ever be executed.

Our experiences and observations indicate that the lack of acceptance of excellence constitutes a strong barrier that has to be overcome for successful innovation.

Excellence is not in good currency. Excellence is only reluctantly accepted in our society. When Mohammed Ali proclaims, 'I am the greatest!', the resentment about such lack of modesty ultimately leads to putting him in prison and robbing him of his title. When Johan Cruijff claims to be recognised as the world's best football expert, he is expelled from the board of his club Ajax. To claim excellence, of individuals as well as organisations, is a taboo.

As Peter Drucker states: *Innovation is work* (Drucker 1985, p. 126). Excellence is acquired by hard work over a great number of years.

If we want to stimulate innovation, we should appreciate and reward those who display excellence, not envy them.

L. A. van Gunsteren, A. G. Vlas, *The License Giver Business Concept of Technological Innovation*, Future of Business and Finance,
https://doi.org/10.1007/978-3-030-91123-2

Appendix A: Planning for Technology as a Corporate Resource: A Strategic Classification[1]

This paper describes a classification of business identities based on the double dichotomy of doing vs. thinking and product vs. capacity. Successful business units, in the sense of social units having their own style and shared values, tend to fit into one of the four basic identities: License Giver, License Taker, Jobber or Consultant. Conversely, a mixture of these in one unit tends to lead to a lack of clear identity organisational stress and lack of strategic impact. Related organisational dilemmas can be resolved by restructuring the organisation according to the four basic identities of the strategic classification.

Introduction

Modern concepts of strategic planning tend to emphasise the need to integrate, in a balanced way, the strategy and the structure with the culture of an organisation. The selection of strategy and structure is to a great extent at the discretion of the management. The culture (the set of norms and values determining 'how things are done over here') has, on the contrary, largely to be considered as given, since deeply rooted beliefs are extremely difficult to change. The organisation's culture is related to the organisation's identity, being the answer to the question 'who are we?'

That identity can relate to many aspects, e.g. is our orientation long term or short term, national or international, etc. of which, in our era of technological change, a most important question is how does the organisation deal with technology?

Technology is a resource that one can buy and sell as well as generate. Which of these approaches is predominantly embedded in the organisation's culture largely determines the appropriate choice of both strategy and structure. If that choice is

---

[1] Long Range Planning, Vol. 20, No. 2, pp. 51–60, 1987.

vague or poorly matched with the culture, organisational stress and lack of strategic impact must be expected.

In this article we propose a classification of strategies related to the use of technology as a corporate resource which will allow us to assess if the organisational culture on the one hand and strategy and structure on the other are in harmony with each other and to identify recommendable changes in the latter two wherever deemed necessary.

Exploitation of R&D Output: Four Typical Cases

A well-known rule of thumb indicates that roughly only one out of every ten projects completed by the R&D laboratory becomes a commercial success. Apparently, the proper exploitation of R&D output constitutes a major problem for most organisations. Although many reasons can be given for this, the most frequent one is undoubtedly a mismatch between the R&D output and the identity of the organisational unit entrusted with the commercialisation of that output, or alternatively, confusion about the true identity of that unit.

The classification of strategies presented here has proven its practical value in the analysis of such problems. Before explaining it, let us consider four typical cases where the exploitation of R&D output appeared to be a problem.

Case 1

Division A was one of the eight divisions of a corporation with an impressive track record of technical achievements. The division produced gas turbines made to their own designs. Considerable investments were being made in the development of the product's next generation which was expected to be superior in fuel consumption to any existing turbine. It then happened that the largest order in the market over the last 3 years was negotiated against tough international competition.

Ultimately however the President and the Chairman of the Board decided to let the order go to the competition since it could only be produced at a great loss in Division A, although at a still positive contribution to overheads. As a result of missing this key order the company had to lay off workers not only in Division A, but also in Division B which would have been a major subcontractor for the manufacture of parts. The event made it clear that the company could no longer fund the development of the new generation entirely on its own and a decision had to be taken to merge Division A in a joint venture with a leading gas turbine manufacturer.

Case 2

Division X, responsible for the production and sales of controllable pitch propellers, was the problem child of a worldwide manufacturer of marine propellers. Its sister division Y, producer of Monobloc propellers, had so far served as the cash cow from which the losses were funded. Division X was licensee of three different designs, but wished to develop its own design for which purpose, in spite of the envisaged loss-making situation, substantial investments for development were made. The output of the development effort, however seemed to be again and again too little and too late; the leading competitor had already twice managed to fill the gap in the market well before Division X was ready. What could the management do about this?

Case 3

Division P manufactured several kinds of electrical equipment. The division originated from the workshop of the company's main line of business being electrical installation contracting which was taken care of by Division Q. For many years losses in Division P were compensated by Division Q's profits. The strategy adopted by the management of Division P to make the division profitable was to develop and market new products which were supposed to gradually take over production capacity that initially was used for subcontracted, 'jobbing' work for Division Q and third parties. After some years one could observe:

1. More than 70% of the production was still jobbing work.
2. Many new products had been launched but in respect of both volume sold and profitability all had failed.

Two products were particularly illustrative of the situation.

1. Emergency illumination. Initially, this new product was a success. Substantial numbers of it were sold by sister Division Q. Eventually, however, other manufacturers took over the idea and brought cheaper versions on the market. Ultimately even Division Q had to buy from others if they were not to impair their position in their own trade.
2. Electronic organ. As a by-product of all kinds of electronics-related development work a new type of electronic organ was invented. The prototype showed, according to experienced musicians, several advantages in comparison with existing types. Nevertheless, not a single one was sold because the company had absolutely no access to the essential distribution channels. Electronic organs are sold via shops of musical instruments and not via shops of electronic gadgets.

How could it happen that in spite of genuine effort on the part of management and substantial investments the strategy of engaging in new products produced, quite contrary to the intention, only losses?

Case 4

The R&D department of an international contracting firm had developed an apparatus for subsea soil investigation. The equipment was capable of carrying out soil investigations in depths of up to 6 m in the sea bed at maximum 200 m water depth. It could perform three functions: drilling, sampling and Dutch cone penetration. Its use was primarily intended for soil investigation for dredging operations but it could also be used in various offshore applications like foundations for offshore constructions and projection of subsea pipelines. After its first successful application a policy for its further commercialisation had to be developed. Three options were open to management:

1. Give exclusive rights to the corporation's dredging company
2. Give the rights of exploitation to the firm's survey and soil investigations company (actually acting more in the capacity of an engineering consulting company) or
3. Give the rights of sale and production to a hardware manufacturer.

Finally, a mixture of options 1 and 2 was chosen: the survey company would exploit the equipment by hiring it out to the dredging company as well as to third parties but the dredging company would have a veto right in respect of its use by competitors. During the following years the survey company was able to profitably hire out the equipment and also used it as a leverage to sell engineering. However, the improved Mark II and Mark III models which the development team had hoped for were not produced and the equipment's great potential for the offshore industry (it could drill a hole in the sea floor at one tenth of the costs involved when using a manned diving bell) was never realised. Subsequently, after 3 years and thorough intelligence work competitors produced their own planning equipment based on the same concept. How could it happen that no more advantage could have been taken of the technological edge which the company had?

Classification of Strategy

These and numerous other cases have led us to the conclusion that to establish the identity of an organisation the following two questions are of particular importance:

1. Are we an organisation of doers or thinkers? In other words are we in a business of making or doing things, or are we in a knowledge business?
2. 'Are we offering a product or a capacity to our customers?'

The importance of the latter distinction has been stressed by Simon, in his analysis of manufacturing organisations in The Netherlands. The four possible combinations of answers to the questions can be placed in a matrix (Fig. A.1).

Fig. A.1 Classification of business identities

	Doing / Making	Thinking / Knowing
Product	**License Taker**	**License Giver**
Capacity	**Jobber** (High or Low Technology)	**Consultant**

We have labelled the four quadrants:

- License Giver
- License Taker
- Jobber (high or low technology)
- Consultant

License has to be taken here in the broadest sense of the word. A License Giver may not actually give licenses, or even take them on certain components or subsystems. The essence is that its raison d'etre is to generate new knowledge related to a particular product.

The classification of Fig. A.1 has proved useful in discussing strategic issues like design leadership and geographical market penetration but also the required management profiles, the organisational culture and the requirements that have to be met by the accounting function of a business unit.

Close observation of companies has led us to the conclusion that successful firms tend to fit in just one of the four quadrants or have separated their organisational sub-units in such a way that each one clearly fits into only one quadrant.

Strategic dilemmas and organisational stress tend to occur when the different characteristics associated with each of the four business identities simultaneously appear within one organisational unit.

We will now explain the nature and describe the characteristics of the four basic identities.

1. License Giver

Figure A.2 represents the revenues and expenditures of a market leader in the manufacture of ship engines. The bulk of the revenues come from license fees and only a minor part from own production. The cost price of the engines produced in the corporation's own production facilities is about twice the cost level of its major licensees.

The own production of engines is nevertheless continued because of the need for direct feedback from the field which is vital to develop the next generation of the product. When the life cycle of the current type has expired the new type should not only incorporate new technologies which have become available but should also comply with changes in requirements by end users. The latter is only possible when a continuous feedback from operations is provided. The 'loss' on production should therefore be seen as a special kind of development cost; together with the R&D costs they constitute the RD&D costs, i.e. the costs of Research Development and *Demonstration*.

The typical product life cycle (Fig. A.3) and the costs of development and demonstration of a new version are determining factors for the annual amount which has to be spent on RD&D in order to remain in the race as a License Giver.

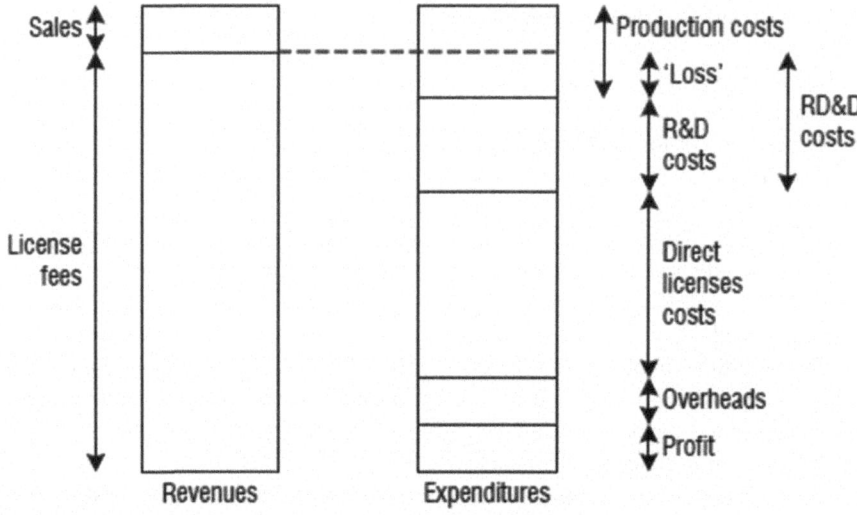

Notes: • 'Loss' on own production should be considered
to be costs of **prototype** development
• Important means to control licensees:
keep production of one vital part in your own hands

Fig. A.2 Annual revenues/costs structure of extreme type of License Giver

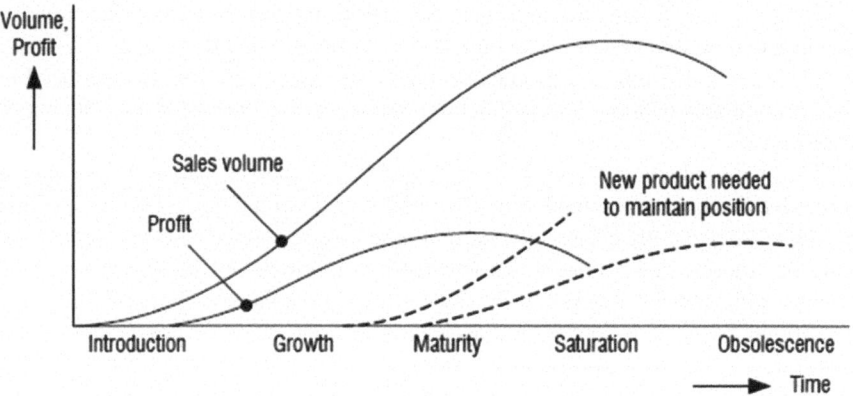

Fig. A.3 Product life cycle

The amount is calculated as follows:

• Typical life cycle—L years
• Cost of development—X million
• Cost of demonstration—Y million

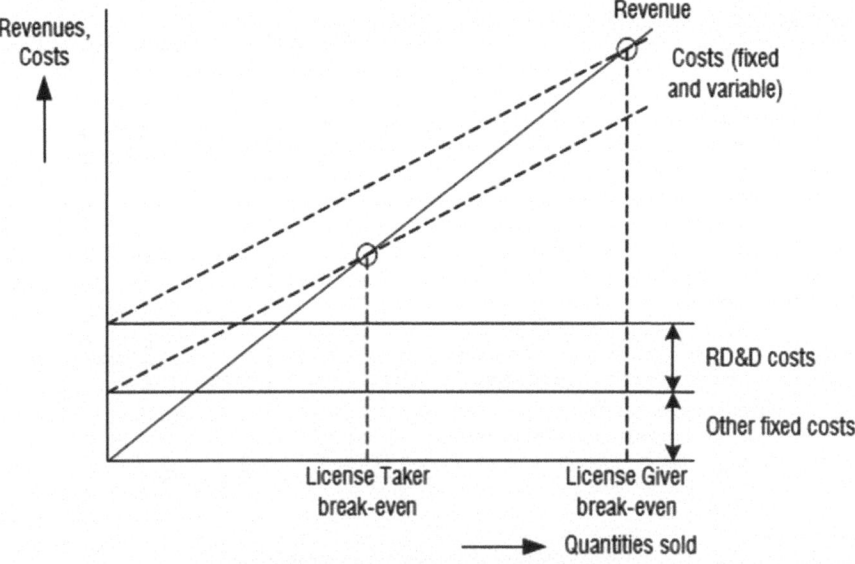

RD&D costs: Average yearly RD&D (fixed) expenditure to
develop a new prototype at a frequency life
cycle typical for the product concerned

Fig. A.4 Importance of market share to afford R&D

so required RD&D expenditure is $(X + Y)/L$ per annum.

If we spend less, we will fall short of critical mass and the new product will arrive too late on the market. This does not imply, of course, that spending the right amount will result in guaranteed success.

These RD&D expenditures are to be recovered from sales and license fees. As a result the License Giver break-even point is positioned at a much higher turnover level (in numbers sold or licensed) than in the case of a License Taker (Fig. A.4).

When the own production becomes small in comparison with the licensed production, the profitability of it becomes of secondary importance. The own production is then only maintained as a means to receive direct operational feedback.

The break-even chart of the extreme type of License Giver is shown in Fig. A.5.

In short, strategically the name of the game for the License Giver is to get as many as possible of his products placed onto the world market. The typical sequence to achieve this is

1. Secure home market
2. Export via agent
3. Export via local sales office
4. Form local joint venture
5. Arrange local license taker (production and sales)

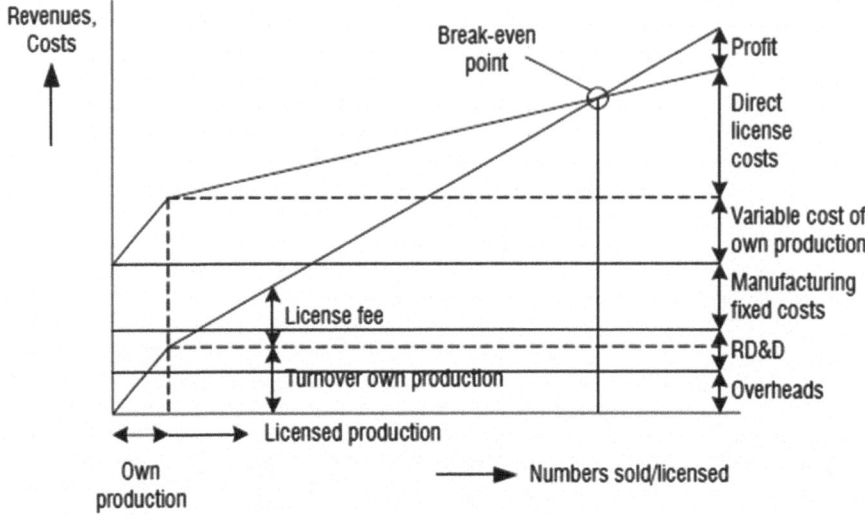

Fig. A.5 Annual cost—revenue structure of extreme type License Giver

The latter two are necessary to overcome protectionism.

In general, managers of product divisions of a corporation tend to prefer export via agents or sales offices but are reluctant to engage in J.V.'s or full license giving because of the perceived loss of control. From a strategic point of view however, if the company is not to lose its design leadership it is essential to make the transition to these stages in time. A powerful means to maintain control over licensees is to exclude one vital (patented) part from the license agreement. The licensee is then forced to purchase that part from the License Giver who by this means keeps in touch with the actions of the licensee in the market. The features of the License Giver are summarised in Table A.1.

In the strategy classifications of Ansoff and Stewart (Ansoff and Stewart 1967; Blake 1978) the license giver pursues a 'First-to-the-market' or a 'Follow-the-leader' strategy.

The emphasis of the License Giver in operations and accounting is summarised in Table A.2.

2. License Taker

The License Taker's aim is to exploit the potential of a particular existing product in a limited regional market. The features of the License Taker are summarised in Table A.3.

In the classification of Ansoff and Stewart the License Taker pursues a 'Me too' or an 'Application engineering' strategy. The emphasis of the License Taker in operations and accounting is summarised in Table A.4.

3. Technology Lobber

The (high or low) Technology Jobber offers a multi-functional manufacturing or servicing capacity to the local market. His main concern is to get his capacity utilised to the full (Fig. A.6).

Table A.1 Features of License Giver

> · Key word: design leadership of product (worldwide)
> · Own production provides operational feedback to develop next generation
> · High overheads (extensive RD & D)
> · In principle, worldwide outlets
> · Management orientation focused on:
>   o maintaining design leadership (effectiveness rather than efficiency; benefit consciousness rather than cost consciousness)
>   o fostering an innovative, entrepreneurial climate
>
> · Sales taken care of at middle management level
> · Subcontracting as much as possible
> · Financing primarily for new product development and demonstration
> · Decision making dominated by long-term strategic considerations
> · Short-term pricing decisions heavily influenced by direct costing considerations

The features of the Technology Jobber are summarised in Table A.5.

The 'stuck-in-the-middle' organisation is bound to fail due to the lack of clear identity (Fig. A.7).

In his book on competitive strategy Porter (Porter 1980) also warns against a lack of choice between the three generic strategies he defines (Tables A.6, A.7 and A.8).

The firm which fails to develop its strategy in at least one of these three directions (a firm which is 'Stuck in the Middle') is in an extremely poor strategic position (Porter 1980). Porter's generic strategy of differentiation comes close to the License Giver of our classification but no distinction is made between a product firm and a capacity firm. Although they both may pursue a generic strategy of overall cost leadership, as previously explained, their nature is fundamentally different. The issue of dedicated versus general purpose facilities is a fundamental one having an impact on almost every aspect of strategic management of a business.

In his book, Competitive Strategy, Porter also warns against a lack of choice between the three generic strategies he defines: Overall cost leadership and Differentiation which are industry wide and Focus in a particular segment only.

The firm which fails to develop its strategy in at least one of these three directions, a firm which is 'stuck-in-the-middle', is in an extremely poor strategic position (Porter 1980, p. 411). Porter's generic strategy of Differentiation comes close

Table A.2 Emphasis of License Giver in operations and accounting

1. Product design
   - estimation of prototype production costs
   - estimation of market potential
   - establishment of life cycle
   - establishment of schedule for efficient production (for License Taker)
   - estimation, and periodic review resulting in revision of estimates of standard production costs (also for License Takers)
   - incorporation of (updated) field feed-back into prototype design and (revised) cost estimates
2. Own production
   - scheduling (short assembly time is essential)
   - quality control (if one sub-system fails the whole system fails)
   - cost control with main emphasis on purchasing costs

to the License Giver of our classification but no distinction is made between a product firm and a capacity firm.

Although they both may pursue a generic strategy of overall cost leadership, as previously explained, their nature is fundamentally different. The issue of dedicated vs general-purpose facilities is a fundamental one having an impact on almost every aspect of strategic management of a business (Table A.9).

Table A.3 Features of the License Taker

- Key word: efficiency
- Outlets in limited regional market (sales and production both local)
- Emphasis on technical development is on process
- technology (to keep production costs low) and on custom engineering (to adapt the product to local market requirements)
- Moderate overheads (mainly in the areas of sales and service)
- Management orientation focused on:
  - regional aspects, for instance relations with key customers, labour unions and local government
  - cost consciousness
  - medium-term horizon (moderate risk and moderate profitability)
  - fostering thoroughness and discipline
- Sales taken care of at middle management level
- Subcontracting as much as possible
- Financing primarily needed for replacement and extension of production facilities (tailored to product)
- Short-term pricing decisions heavily influenced by local market conditions
- Emphasis in accounting on standard costs and purchasing (goods and services)

Table A.4 Emphasis of the License Taker in operations and accounting

1. Purchasing on call orders
2. Production for inventory (large series)
   - scheduling geared to maximum efficiency
   - cost control through standard costs with variance analysis
   - inventory cost control
   - cost control of workforce (shifts)
3. Production on customer orders (small numbers)
   - critical path analysis to meet delivery time
   - overall cost control through project cost control (cost estimate => progress => estimate to complete alternative critical path => revised estimate => etc.)
   - squeeze on his 'technology jobber' (sub-supplier)
   - control on efficiency per department by standard departmental shop floor cost (variance analysis)
   - control on material cost by competitive bids governed by quality specifications and track record of timely delivery (reliability reputation of sub-supplier)

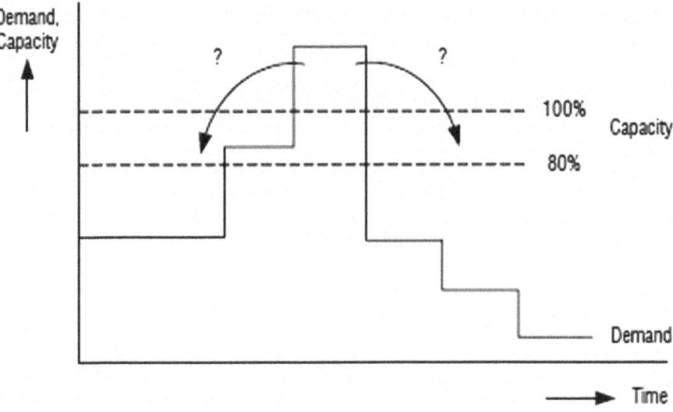

Fig. A.6 Levelling problem of the Jobber

Table A.5 Features of the Technology Jobber

> - Key word: occupancy of multi-functional utility (general purpose facility)
> - Outlets regionally limited by tariff barriers and transportation
> - Emphasis of development, if any, on process technology
> - Low overheads
> - Management orientation focused on:
>   - plant utilization
>   - cost consciousness
>   - short-term horizon
>   - flexibility and labour motivation ('we'll fix it' mentality)
> - Sales taken care of at highest management level (knowledge about deadline exposures of current and potential customers)
> - Subcontracting as little as possible
> - Financing primarily needed for replacement and extension of production facilities (general purpose)
> - Decision making, including short-term pricing policies, dominated by planning of plant utilization

Fig. A.7 Stuck-in-the-Middle identity crisis

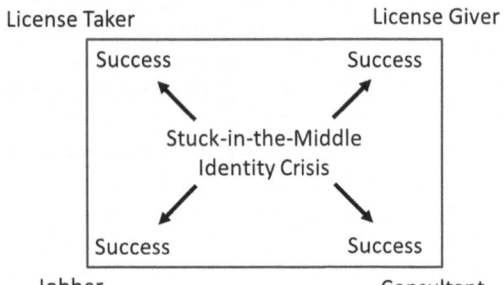

Table A.6 Emphasis of the Technology Jobber in operations and accounting

1. Emphasis on capacity efficiency; accounting system should provide the basis for discounts to customer for rescheduling (disproportion in area above 80% should be readjusted by means of discounts and extra charges, see Figure 6)
2. Post-mortem review to establish unit or job cost
3. Where 'main-supplier': control on job cost by exploiting learning curve
4. Control of efficiency per department by standard departmental shop floor cost (variance analysis)

Table A.7 Features of the Consultant

- Key word: Customer service
- Hiring out of knowledge capacity
- Emphasis of development activities: incorporation of new techniques in procedures and programmes
- Low front investments
- Low overheads
- Management orientation focused on:
  o selling man-hours, mainly with a short-term horizon
  o productivity
  o development of consultant's skill (to maintain level of knowledge)
- Selling at all levels, however main sales taken care of at highest management level
- Subcontracting as little as possible
- Decision making tends to be opportunistic

Table A.8 Emphasis of the Consultant in operations and accounting

> 1. Utilisation of available hours (minimum 75% of 1500 hours per annum per person)
> 2. Budgeted man-hours versus actual
> 3. Stringent cost control on overheads
> 4. Budgeted training costs as a percentage of standard per diem rates
> 5. Stringent cost control on out-of-pocket expenses (authorisation at highest management level)

Table A.9 Comparison of the features of the four basic identities

	License Giver	License Taker	Jobber	Consultant
Name of the game	Largest possible number on the world market	Capture of regional market by favourable price/performance ratio	Plant occupancy	Utilisation of (knowledge) man-power
R&D emphasis	Product design leadership	Process technology	Custom engineering	Adapting available techniques to customer needs
Time horizon	Long term	Medium term	Short term	Short term
Geographical focus	World	Country	Local	Local
Organisation climate	Innovative, Entrepreneurial	Discipline	Flexibility, labour motivation	Opportunistic
Cost emphasis	Effectiveness, Benefit consciousness	Efficiency, Cost consciousness	Cost consciousness	Out-of-pocket expenses and overheads
Overheads	High	Moderate	Low	Low
Sales	At middle management level	At middle management level	At first management level	At all management levels
Subcontracting	As much as possible	As much as possible	As little as possible	As little as possible
Main thrust of investments	RD&D	Dedicated plant	General purpose equipment	Training

The Multi-business Corporation

The basic identities can flourish alongside each other within one corporation, as long as the associated organisational units are kept separate and their autonomy is sufficient to allow them to develop their own appropriate approach to business problems (along with a general corporate spirit). Conversely, when the identities are mixed up in one organisational unit a split-up will resolve most of the prevailing organisational dilemmas. Such a split-up introduces the problem of intra-corporate deliveries. More often than not we see that one unit is either obliged to purchase from the sister unit or to give it at least the right of first refusal. This is a misleading concept which should be avoided. Third parties will very soon find out that they are only used as a price leverage to bring the sister unit to a lower price but that they never get any orders.

The result is that they do not make a serious offer or they straightforwardly ask the sister unit what price they should quote and the whole procedure becomes a ritual. An effective means to induce genuine competition without having the well-known drawbacks of general comparative shopping is the *concept of the second main supplier*: for each strategically important purchasing segment a firm should see to it that it gets (at least) two main suppliers. The second main supplier should get, over a longer period of time, at least 30%, the first one a maximum of 70% of the relevant purchasing segment. This means that the second main supplier should temporarily get a right of first refusal whenever the balance has to be restored. In this way both suppliers will remain alert which is not only in the interest of the purchasing unit but will also improve the competitive strength of the suppliers (including the sister unit) in the open market.

Discussion of Four Typical Cases

Let us now return to the four cases cited at the beginning of this article and see how they fit in our scheme.

Case 1

Division A, producing gas turbines according to their own design, is a clear cut case of a License Giver. A major part of the corporation, however, could be characterised as a High Technology Jobber, which apparently heavily influenced the decision-making by corporate management. Post-mortem analysis of the key order concerned revealed the following:

- Plant utilisation, i.e. the key issue of the Technology Jobber, was a major consideration in top management decision-making.
- The 'walk-out' price that corporate management had in mind was entirely based on the contribution that would be generated in Division A; that a substantial contribution to overheads would be realised in Division B was completely overlooked as a result of insufficient insight into the transfer pricing procedure those geographical areas where Division X was (Fig. A.8).

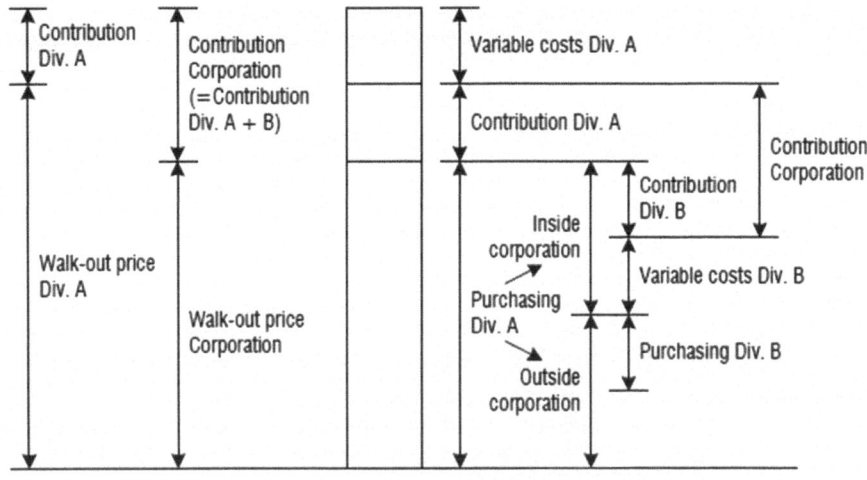

Division A: License Giver
Division B: (High-tech) Jobber
Contribution: fixed costs, overhead, interest, profit

Fig. A.8 'Walk-out' price can be significantly affected by interdivisional deliveries

- Letting this key order go to the competition actually meant ceasing to be a real License Giver; this fact was only realised a considerable time after the decision had been taken; as a result commensurate measures to cope with it were taken at a slow, and heavily loss-making, pace.

Case 2

Division X being licensee for three different designs in addition to development efforts related to an own design was actually in a 'stuck-in-the-middle' position between License Taker and License Giver. Break-even analysis (Fig. A.5) revealed that the company either had to return to its position of pure License Taker or complete its transition towards a position of License Giver. To achieve the latter the turnover in terms of numbers (of controllable pitch propeller installations of own design) had to be doubled. A marketing strategy to this end was devised and implemented. The turnover was doubled in 2 years. The marketing strategy included the following features:

1. Stressing reliability of the (own) design; reliability of the propeller is directly related to the availability of the whole ship and therefore the key selling function.
2. Direct marketing efforts towards ship-owners, i.e. the end users, rather than towards shipyards.
3. Using the concept of nuisance value vis-a-vis the competitor, i.e. displaying marketing efforts in his home market. Consequently, if he wished to maintain the

price level in his own home market he had to respect price levels in those geographical areas where Division X was strong and in a position to increase market share.

4. Regionally *differentiated price* levels.
5. Progressive and regionally *differentiated commissions* to agents.

The Division thus definitely succeeded in establishing itself as a License Giver and became the main profit maker of the corporation.

Case 3

As Division P originated from a workshop for the installation division it possessed the typical features of a Jobber. Its manager however, wishing to make the Division a manufacturer in its own right, emphasised the development of own products. That is to say, in words and not in behaviour which remained focused on satisfying short-term needs of customers he happened to be in contact with. The short-term problems always took precedence over the longer-term opportunities. As a result jobbing work remained the main source of income which was not sufficient to compensate for the substantial losses on the own products. A strategy to return to a pure Jobber status was therefore adopted and profitability was gradually restored.

Case 4

The apparatus for subsea soil investigation had the potential of entering the market of subsea equipment as a License Giver. Entrusting its commercialisation to an Engineering Consultant made this no longer possible. The management used its unique selling points to sell engineering hours and thereby failed to exploit its longer-term potential by sustained development of a Mark II and III. From their point of view as a Consultant that was perfectly in order, but the key people involved in the development left to join License Giver type companies in subsea equipment.

Practical Implications

To summarise, adopting a strategy as to how to use technology as a corporate resource should imply a choice which should be in line with the existing identity of the organisation. The choice should be deliberate and not be merely going for the only remaining option. An RD&D strategy not only implies a choice as to what to pursue but also as to what not to do. When these points are neglected R&D output cannot be properly commercialised or it will lack critical mass as a result of splintered effort.

In practice this includes four distinct steps:

1. Analyse the history of the firm in terms of its realized strategy, i.e. what was actually done (for a discussion on the distinction between intended, emerging, and realized strategy see Mintzberg 1978). This allows characterisation of the

way of thinking of the dominant coalition of the firm in terms of License Giver, License Taker, Jobber or Consultant.

2. Identify activities which do not fit in the business identity of the dominate coalition as found in step 1.
3. Revise the organisation structure in the sense that the activities identified in step 2 are separated in units having the specific character according to one of the four basic business identities. Sufficient autonomy must be given to these units to allow them to develop their own style of doing business which will necessarily be different from that of the dominant coalition.
4. Let the business units as defined in steps 1 and 3 develop their own business strategy.

An example may help illustrate this. A firm engaged in both publishing and printing having these activities organised in a highly intertwined way found itself in a continuous loss-making situation. The publishing part of the organisation was held responsible for this; the common opinion in the company was that the book titles were not of sufficient quality to be attractive to the public. A publishing business is by our classification a License Giver, whereas a printing business is a Jobber.

A reorganisation was therefore carried out which was directed to separating the firm into fairly autonomous units. The Publishing unit was organised in line with the typical License Giver characteristics. Contrary to past practice, it was allowed to have books printed by third parties. The printing unit was organised in line with the typical Jobber characteristics. The utilisation of its capacity was made their own responsibility in the sense that they had to regard the publishing unit as one of their customers and no longer as a scapegoat of their own problems. As a result buck-passing and internal quarrelling came to an end and profitability was restored.

References

M. Simon, Kontinuiteitsanalyse van indust-tiele bedrijven, Bedrijfskunde, 52 (1). (2) (1980).

Ii. I. Ansoff and John M. Stewart, Strategies for a technology based business, Harvard Business Review, November-December, 45 (6). 71-83 (1967).

S. P. Blake, Managing for Responsive Research and Development, Freeman, San Francisco (1978).

Michael E. Porter, Competitive Strategy-Techniques for Analyzing Industries and Competitors, The Free Press (1980).

H. Mintzberg, Patterns in strategy formation, Management Science, 24 (9) May (1978).

Appendix B: Use of the Quality Classification in the Construction Industry[1]

A construction company is a Jobber, delivering a capacity, not a product, to build something. Its strategic quality aim is to realise the subjective wishes of the customer. In general, these wishes are as follows:

1. Functionality, delivering something that works as intended
2. Commissioning on time
3. Cost within budget

This means that one should not focus on getting execution in line with specifications, rules and regulations, but on letting execution cover as closely as possible relevant quality as required for functionality. We describe how the classification of the seven categories of quality was applied in the Nanhai project, a US$ 4.3 billion construction project of a petrochemical plant in the Guangdong Provence of P.R. China (Van Gunsteren 2011).

Prerequisite for the Implementation of any New Concept: The Product Champion

For the construction industry, the principle that compliance to specifications should be subordinated to real quality, i.e. fitness for purpose, constitutes a new concept. New ideas do not sell themselves. They need a product champion, also called organisational guerrilla, to achieve acceptance. The product champion fights with all available means for the acceptance of the innovation and is prepared to risk his reputation or even his job for it. A product champion is a prerequisite for the

---

[1] Chapter 7, pages 51–56, from "Quality in Design and Execution of Engineering Practice", 2013 L.A. van Gunsteren, IOS Press.

© The Author(s), under exclusive license to Springer Nature Switzerland AG 2022
L. A. van Gunsteren, A. G. Vlas, *The License Giver Business Concept of Technological Innovation*, Future of Business and Finance,
https://doi.org/10.1007/978-3-030-91123-2

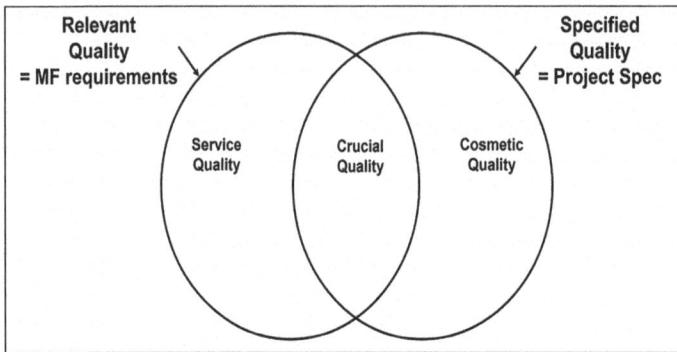

Fig. B.1 Specs never cover relevant quality

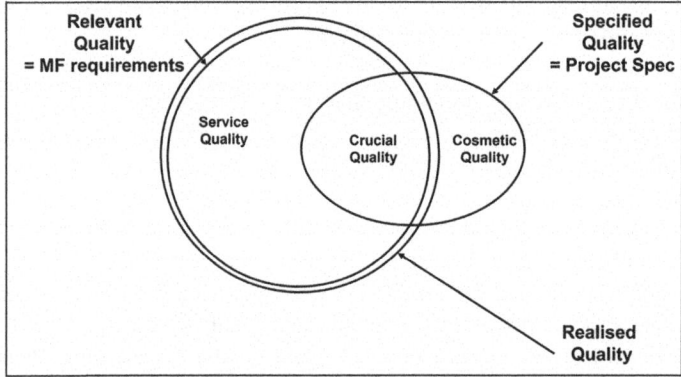

Fig. B.2 The ideal world

implementation of anything new in order to overcome the fear of innovation, which prevails in every organisation.

Fortunately, our quality concept found its product champion in the person of Ton Sluman, who understood it and applied it in his daily work. He attached the pictures of the circles with associated one-liners on the publication board on the site (see five posters, Figs. B.1, B.2, B.3, B.4, and B.5).

Support for the product champion's approach was provided not only by the project team but also by the CEO of the entire project (Simon Lam). The prerequisite for acceptance of a new order of things, a product champion with the blessing of a benefactor high up in the organisation, being satisfied in this case has been a key factor for its success.

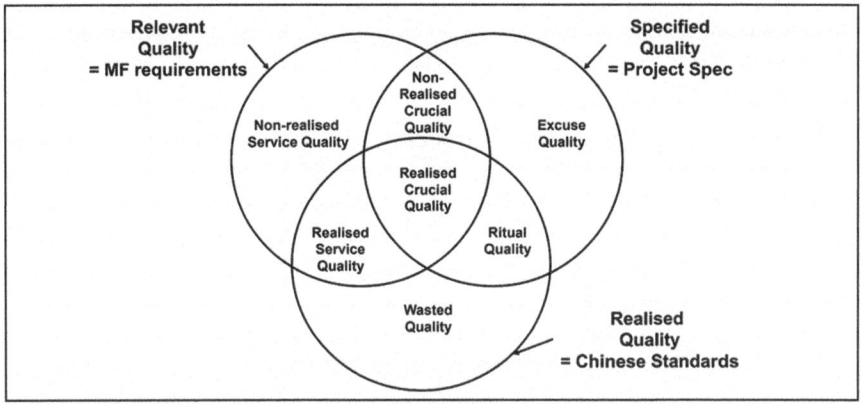

Fig. B.3 You do not get what you want or specified

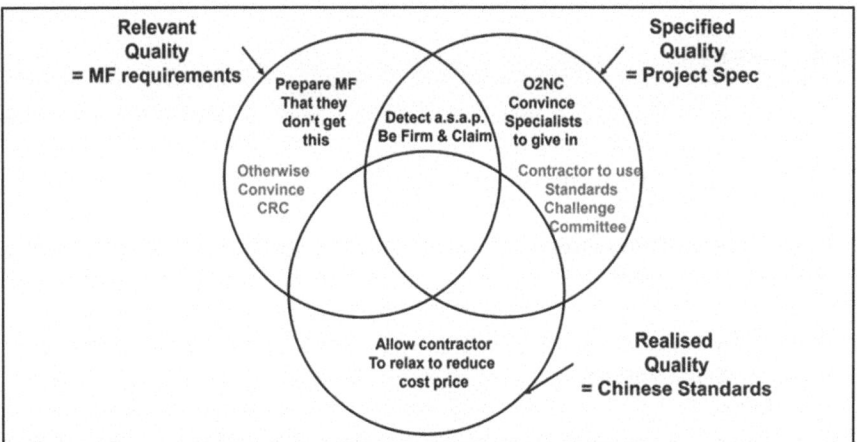

Fig. B.4 Communication is key

The Use of the Quality Circles

The Venn diagrams of our quality classification, usually referred to as quality circles, were used for three different purposes:

1. Accepting Chinese standards whenever possible. For instance, accepting them for buildings not essential to the process of the plant, and relaxing the offshore Shell standards to Chinese ones for the jetty in the middle of Daya Bay.
2. Changing scope. Many small changes done in process to increase reliability and reduce costs. For instance, delaying the railway, since it could not yet be used by the future refinery next door and outsourcing the air splitter to Praxair adjacent to the site, who could make also nitrogen and oxygen for others, thereby utilising economies of scale to the benefit of both parties.

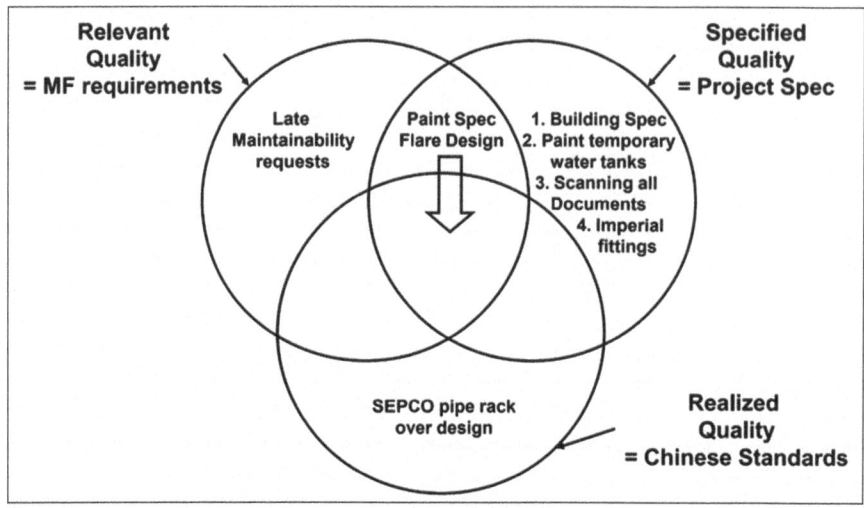

Fig. B.5 Examples

3. Managing expectations of maintenance departments by providing them with records of non-conformance from original specifications for later use in inspection programs and debottlenecking studies.

Quality-related issues arose in all seven categories of quality, as is illustrated in the following examples:

1. Non-realised service quality (relevant, not specified, not realised). Hidden deficiencies which surface in the first commissioning phase and first years of operation.
2. Non-realised crucial quality (relevant, specified, not realised). Serious issues arose with the quality of underground water cooling lines due to sub-standard design by Chinese vendors, irregularities with licenses, and construction not according to specifications but to Chinese practices of drains. Awarding contract after competitive tendering to four different contractors entailed losing central control. Ultimately, a fallback system (based on steel instead of glass fibre reinforced Epoxy) was installed, but so far has not been used.
3. Realised service quality (relevant, not specified, realised). Example 1: Because of accidents elsewhere, spheres for ethylene and propylene storage were under scrutiny. German materials were used which were better than prescribed by the Chinese authority for this matter. Nevertheless, a lobbying battle turned out to be needed for their approval.

    Example 2: Dredging by Chinese contractors without dumping. Monitoring for suspended soils and other environmental impact done by Boskalis was better than specified. Even a living choral was relocated. But Chinese authorities did not believe there had not been any dumping, invoked the license and wanted penalty fees to be paid. Instead of giving in to this, a budget was approved for additional inspection at a dumping station, 20 miles out of the coast in accordance with the London dumping convention. Apparently, it is so unusual to do better than specified in regulations that this evoked problems instead of appreciation.

4. Wasted quality (not relevant, not specified, realised). Overdesign in engineering is quite common in China. The government holds engineering, which is kept completely separated from construction, responsible for eventual disasters. Chief Engineers of Chinese design institutes, trying to reduce the risk of sanctions (including prison sentence), tend to be conservative rather than cost-conscious. Construction, by contrast, is in China always trying to cut costs by compromising quality. For this reason, the government established specialised supervision companies to check on 'construction to design'. In the Nanhai project, the steel constructions for the power plant were overdesigned by Sepco, a Chinese power and construction company.

5. Ritual quality (not relevant, specified, realised). A big investment was made in the water treatment and solid waste disposal facilities. A 20 km pipeline was laid under water to discharge at a point of maximum turbulence and mixing with tidal movements. Other local parties in similar situations refrained from such expensive measures, indicating that this was a case of ritual quality.

6. Realised crucial quality (relevant, specified, realised). Many issues surfaced in this category. One example is small bore connections: Chinese contractors were not sufficiently aware of the specifications and did not comply. Corrections were made on time with special inspection tools. A second example is flare construction, which was a copy of the plant in Pernis which can be lowered during operation. This requires sliding tolerances in millimetres. The Chinese vendor, not being sufficiently warned on this point, produced power tower quality with tolerances in centimetres. It was redone at the site, on time but at extra cost.

For the authorisation of the quality-related changes, two committees were in function:

1. Change committee on scope changes, chaired by the CEO (Simon Lam).
2. Standard challenge committee, chaired by the construction director (Frans van Gunsteren), which always involved the end user in its decision-making.

Giving away cosmetic quality and wasted quality, or exchanging these for service quality, yields substantial cost savings for the contractors concerned and generates valuable goodwill with them. It is essential however that the changes are authorised at the right organisational level.

Trade-Offs Between Quality, Costs and Schedule

In construction projects, trade-offs must always be made between quality, costs and schedule. Quality is usually perceived as being defined by the project's specifications. Costs are supposed to be specified by the budget. The schedule is assumed to be given by a network planning aimed at completion on time. As a result, prevailing management focus in construction projects tends to be concentrated on cost and schedule, with quality management limited to implementing contractual specifications.

When construction projects become large and complex, however, many relevant matters are reflected neither in the contractual specifications nor in the budget or the network planning of the project. As a result, functionality suffers under the prevailing management approach. A costly effort must still be made to ensure that unspecified, yet relevant, quality is realised in the project. Inevitably, this leads to substantial overruns in time and money.

Attempts to avoid these overruns in time and money have resulted in ever more exhaustive specifications, budgets and schedules, but these turn out to produce disappointingly little effect. This is not surprising in view of the fact that the every-day reality is too complex to be realistically reflected in specifications, budgets and schedules. Even if that would be at all possible, it is naive to expect that subcontractors will take the time to fully read and digest such voluminous documentation, particularly within the limited time of the bidding phase.

In short, the usual preoccupation with cost and schedule does not work. As with the arts of Zen – archery, sword fighting, flower arranging – one has to remove the ultimate goal – the arrow hitting the target, striking the opponent, achieving the most beautiful flower arrangement – completely from the mind and concentrate on quality as required by functionality and not only as specified in contracts and consider cost and time of completion as outcomes of a process, which can only be indirectly be controlled.

This requires a lenient management style, different from the control-oriented top-down management style as recommended in the main stream of literature on construction project management. The concept of distinguishing best-management practices PI for simple projects from best-management practices PII for complex projects has been conceived by van Gunsteren and van Loon by means of combining best-management practices from Industrial R&D with those of Open Design methodology aimed at solving the multi-stakeholders design problem in architecture and urban planning (Binnekamp et al, 2006, pp. 109–122).

The concept is condensed here in the table below. PII management is also referred to as *Floating goals management*.

Table B.1 Best-management practices PI for simple and PII for complex projects

	PI (simple projects)	PII (complex projects)
Goal setting	Before awarding contract—for design and/or construction—the design brief or the design itself should be frozen and not be unfrozen before commissioning.	Nothing is fixed in advance, be prepared to adjust goals when circumstances change and insight improves.
Leadership	Leadership is provided by the project manager, who is the central figure in the entire process.	Aim at leadership focused at defending relevant stakeholders' interests.
Conflict resolution	Focus on powerful stakeholders and try to establish compromises between them.	Aim at open synthesis (not closed compromise), i.e. choices aimed at satisfaction of stakeholders concerned.
Design process	Proceed from coarse, preliminary design towards detailed design in a trial and error process starting from an arbitrarily chosen first design.	Proceed from ideal constraints of stakeholders to alleviated constraints to achieve a solution at all.
Communication	Keep everyone involved informed on design status, approved changes and planning.	Respond to information needs and demands of decision-makers (designers, stakeholders).
Persuasion of players	Make presentations to convince players who have to accept compromises.	Persuade by supplying valid and relevant information.
Progress control	Divide the process into small steps with identifiable milestones against planned deadlines.	Pay attention to both 'hard' and 'soft' information on progress.
Division of tasks	Define division of tasks and associated responsibilities in job and function descriptions.	Think in roles rather than tasks, using only broad job descriptions.
Integration and coordination	Integration and coordination of tasks is a prime responsibility of the project manager.	Create a climate for mutual adjustment of tasks.
Standardisation	Standardisation where possible, because standardisation reduces complexity.	Standardisation only where functional and genuinely accepted by stakeholders.

Appendix C: Inductive Thinking, Condition *sine qua non* for Breakthrough Innovation

To explain the difference between deductive and inductive thinking, the first author describes as follows the relevant part of his inaugural lecture: How to deal with technology? A plea for an inductive approach, on November 10, 1982, in the auditorium of Delft University of Technology (van Gunsteren 2004).

We conducted a vote among the audience in the auditorium. The voting offered three options: true, no opinion and false.

The voting procedure had been tested beforehand in the restaurant of the head office of the Royal Boskalis Westminster Group, where I was director of R&D. To ensure that the propositions were understood, some employees of my R&D department were available to give explanation when required.

The results of the voting at Boskalis and the voting during the lecture in the auditorium in Delft are presented together.

An Opinion Poll

How do we generally feel about technology policy and dealing with technology? To this end, eight propositions were submitted to the users of the restaurant of the Royal Boskalis Westminster Group in Sliedrecht. The sample size thereof was 93, and results are presented in the first line. In the second line the results are given of the poll during the lecture in the auditorium in Delft, of which the sample size was 91. The results were as follows:

1. For technical accomplishments it also holds that 'good wine does not need to be praised'.

2. For the exploitation of strategically relevant know-how, one does not need to have technicians on the executive board; it is sufficient to employ them.

true   no opinion   false

3. By subsidising the technology institute TNO (for three-quarters of their budget), the government advances innovation.

true   no opinion   false

4. The classification societies, such as Lloyd's and Det Norske Veritas, are, with their large R&D budgets, an important stimulus for technological innovations in the maritime domain.

true   no opinion   false

5. The strategic potential of a technological innovation project is mainly determinant for success or failure of it.
   Explanation: Strategic potential is made up of four components:

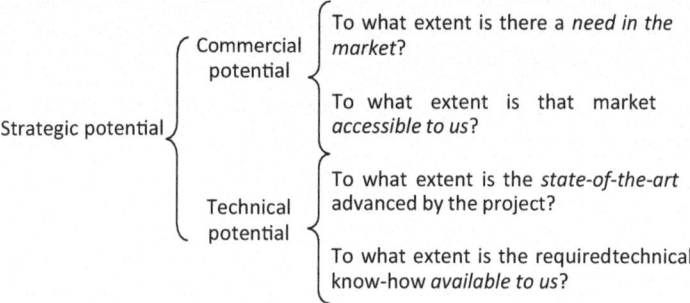

Strategic potential

Commercial potential
- To what extent is there a *need in the market*?
- To what extent is that market *accessible to us*?

Technical potential
- To what extent is the *state-of-the-art* advanced by the project?
- To what extent is the requiredtechnical know-how *available to us*?

If the answer is positive on all four questions, the project has a very high strategic potential.

true	no opinion	false

6. By subsidising R&D in certain priority fields, for instance biotechnology, the Ministry of Economics gives an important stimulus for innovation in the sector concerned.

true	no opinion	false

7. In the science budget of 1982, the following technical sciences are indicated as deserving priority and stimulus (which is reflected in government policies):

- Micro-electronics,
- Robotics and CAD/CAM techniques,
- Information technology (hard- and software),
- Maritime and offshore industry,
- Biotechnology,
- Environment-friendly technology,
- Raw material and energy saving technologies,
- Revalidation research, means for handicapped, medical technology,
- Working conditions,
- Clean technology,
- Oceanography.

By specifying and stimulating these technologies, the government gives direction to technological research in our country.

true	no opinion	false

8. The higher margins in industries with a cartel character often lead to substantial R&D budgets. This situation is conducive to technological innovation.

true	no opinion	false

The conclusion is that all propositions are predominantly perceived to be true. I will show:

1. That this results from *deductive* thinking—that is: general truths are assumed to be valid for special cases;
2. That *inductive* thinking—that is: observing the special case and carefully drawing conclusions from it which might be generalised—leads to concluding that all propositions are false.

Deductive Analysis of Propositions

Deductive reasoning leads to concluding that all eight propositions are true:

1. If, in general, good wine does not need to be praised, why would that not also hold for the special case of technical accomplishments?
2. In general, one does not have to know everything as long as relevant knowledge is accessible. Technical know-how stored in the brains of individuals is accessible when they are on the payroll. Why then would it be necessary to provide them with the formal and sanction power of a director position in addition to their knowledge power (and reference power over R&D workers)?
3. In general, it is true: who pays gets his or her way. If the government subsidises TNO for carrying out their mission to take care of the necessary technical renewal (diffusion of innovation) in the country, it is plausible to expect that this stimulates technological innovation.
4. Mutatis mutandis: if the classification societies spend a lot of money on R&D, isn't it reasonable to expect that this leads to maritime technological innovations?
5. When *demand pull* and *technology push* are both ensured, and our organisation is both technically and commercially well qualified, then success must be guaranteed.
6. Analogue to 3: who pays gets his or her way, at least for a part. Money for priorities must, therefore, stimulate them.
7. Government has both power and money. By defining priority areas and stimulating them with budgets, direction is undoubtedly given to the technical sciences.
8. Analogue to 4: spending a lot on R&D must lead to technical renewal (diffusion of innovation).

Inductive Analysis of Propositions

Inductive reasoning leads to concluding that all eight propositions are false. To this end, we have to pay attention to *observations of special cases* and wonder what these mean for the propositions.

1. The history of the development of technology gives a multitude of examples of valuable inventions that were not accepted initially. The photocopy patent of Carlson, the origin of Rank Xerox, was rejected by twenty companies. Two examples (out of many) from my own field of experience:
   - Only about half a century after the controllable pitch propeller became available, application at some scale came off the ground.
   - The invention of the Kort nozzle dates from before WWII, but its application came only off the ground after the first oil crisis. When applied on a trailer, the saving on fuel costs is about fifteen percent, which brings the payback period within two years, but its application in the dredging industry begins only in 1980.
   - An innovation always brings along a shift in power. Experts in the old technology lose knowledge power to experts in the new technology. It is understandable that they try to abort the embryo when they still are in a position to do so.
2. A necessary condition to exploit strategically relevant technical knowledge is to have someone in the top of the enterprise who has feeling for technology, and in particular a good intuition for what is possible and what is not possible, someone who with understanding of the state of the art in the field concerned. There are two reasons for this:
   (a) Without a technical man in the top with authority, it is unlikely that technology gets the weight required for successful innovation.
   (b) A good timing is essential. Too early or too late leads both to failure. This means that sometimes decisions have to be taken at a stage that the technical man in the top cannot yet prove that his vision is right. His personal authority is then needed to take the necessary measures in time.
3. Several times I have seen that some young engineers abandoned their initiative to start an innovation company because they could not find an answer to my question: 'How do you cope with the competition from the heavily subsidised TNO?'
4. When a shipyard wants to apply something new that deviates from the 'Rules' of the classification societies, they have to expect considerable delays in the approval of drawings. These have to be analysed by technical experts who are higher qualified than needed for usual constructions. The delays can lead to exceeding the delivery time and a penalty for the yard. As long as the classification societies continue to respond to new proposals with delays in their approval, they will remain involuntarily a roadblock to new impulses from the yard or the owner (i.e. the end user).

5. In an inductive study of fourteen cases in the maritime industry, it appeared that the strategic potential and the orientation of top management are equally important for successful innovation.

6. Subsidies for research in engineering sciences are channelled via the STW (Foundation for Engineering Sciences in the Netherlands). About 23% of the STW projects fall in the priority categories of the Ministry of Economics. Let us assume that all projects cost the same, and that a hundred projects are approved. If the Ministry of Economics supplies a modest extra subsidy for, say, five projects within their priority fields, then the STW gets room for supporting five other projects, because double subsidies are out of the question. Statistically, only one out of these five projects will be in one of the priority fields of the Ministry of Economics. If the subsidy is more substantial, say, for twenty projects, those at the bottom of the quality ranking will be of questionable quality, or will not even survive the STW preselection. So, in both cases, the subsidy hardly contributes to the priority fields.

7. The STW uses two criteria for the evaluation of projects: (1) scientific value, (2) utilisation, which means the probability of being applied in practice. The two criteria get equal weight. The procedure, based on referees and independent jury chambers, has been tested on reliability in several ways, statistically and by evaluating certain projects twice, with surprisingly good results. The priority fields of the Science budget of 1982 get their priority on the same criteria (as becomes apparent from the clarification attached to it). As a corollary, we may expect that the percentage of STW projects being within the priority fields to be significantly higher for the approved projects than for the rejected projects. The actual percentages are, however, almost equal. Within the priority fields are: 63% of all projects, 64% of the approved projects, and 62% of the rejected projects. The conclusion is evident. The 'priority fields' comprise two thirds of what makes sense to investigate anyway. Who declares everything to be priority does not prioritise at all. The priority list of the Science budget has no meaning at all.

8. Higher margins resulting from cartel agreements undoubtedly lead to higher R&D budgets. What happens with the results of the R&D efforts is another question. The stability of a cartel can be disturbed when one of its members gets an advantage over the other members by a patented, or hard to copy, invention. The case of the ring propeller described in Chapter 4 illustrates this point.

This does not mean that the propositions are entirely wrong. It shows, however, that dealing with technology is not as simple as at first appears.

What Are We Doing with Technology?

1. Our policy making in regard to technology is dominated by a *deductive approach*. This implies an *arrogance*, which does not recognise the complexity of effectively dealing with technology.

2. The result in regard to science policy is a *fixation on specifying priority fields*. The result in regard to technological company policy is on the one hand a *lack of direction that is in line with the company's identity*, and on the other an abundance of control mechanisms that give only room for *marginal, low-risk innovations*.

What Could We and Should We Be Doing with technology?

1. In the first place, complementing our deductive thinking with an inductive approach: observing what happens and drawing conclusions from that.
2. In regard to science policy this leads to: (1) Initially, no priority fields at all, but leaving to the researchers themselves the choice of their subject.
3. (2) Observing what is well followed up by entrepreneurs. (3) Giving extra support where entrepreneurs and scientists give proof of effective collaboration.
4. In regard to company policy this leads to: (1) Taking technology policy seriously (getting vision in the top, breaking the dominance of short-term thinking, creating an innovative climate including the 'right to fail'). (2) Being prepared to make deeply rooted norms discussable. (3) Being keen on exploiting relevant results of development work and incorporating them in the company's strategies.

This requires a change of our attitude in regard to technology.

Appendix D: The Intrinsically Motivated Crowd of the Eleven Cities Tour, Impressions of a Tour Skater

Revised English translation of the last section of the book *The fifteenth Eleven Cities Tour, heroic battle in icy wind* (1997), by Lex van Gunsteren, who also wrote with Wijnanda Willemse the booklet *Eleven Cities Management, the management of the Eleven Cities Tour and what the business community can learn from it* with a foreword of the chairman of the Eleven Cities Tour Association, Ir. J.S. Sipkema.

When thinking of the 4 Eleven Cities Tours in which I have participated, I realise what appeals to me: 'Like the sea, never the same'.

The Tour of 1963: A frustration for the rest of my life, the harshest tour ever and the last one before the rise of artificial skating rings. Many participants could not skate properly. They leaned too much forward, causing them to stumble on patches of poor ice more often than necessary. Nowadays, most participants have a reasonable skating technique. My brother and I, aged 21 and 25 at the time, skated together. In Stavoren, 66 km from the start, my brother was exhausted and wanted to give up. I refused to continue alone and confirmed his conclusion that his quitting would also spoil it all for me. We agreed to have a rest of 5 min at each city and that I would slow down when he could not keep up. This happened a few times during the track over the IJsselmeer (the former South Sea) where we had to skate against a snowstorm of wind force seven (Beaufort scale). In Harlingen we were told that we had only 35 Tour skaters ahead of us. At 4 o'clock, about 10 h into the tour, we reached Franeker. During our rest, an announcement came through the loudspeakers that the check point in Wier would close at half past 5. All skaters resumed skating at once. Some were even willing to take over the lead from me. We reached Wier, a small village between Franeker and Bartlehiem, at half past 5, just in time to collect the stamp of the secret checkpoint. When we wanted to continue, a policeman ordered us to stop: 'No tour skater will be allowed to continue past this point' he said. 'You must be mistaken', we protested. 'The people at the checkpoint told us that eleven skaters have passed', to which the policeman responded: 'They will be picked up by a motorbike with sidecar'. My brother could not suspect the policeman to be such a straight-faced liar as he turned out to be. Why should we continue when no one else would come farther than we were? This consideration made my brother decide to remove his skates. 'Let's continue till midnight and see where we are at that time',

L. A. van Gunsteren, A. G. Vlas, *The License Giver Business Concept of Technological Innovation*, Future of Business and Finance, https://doi.org/10.1007/978-3-030-91123-2

Fig. D.1 Map of the Eleven Cities Tour track with distances

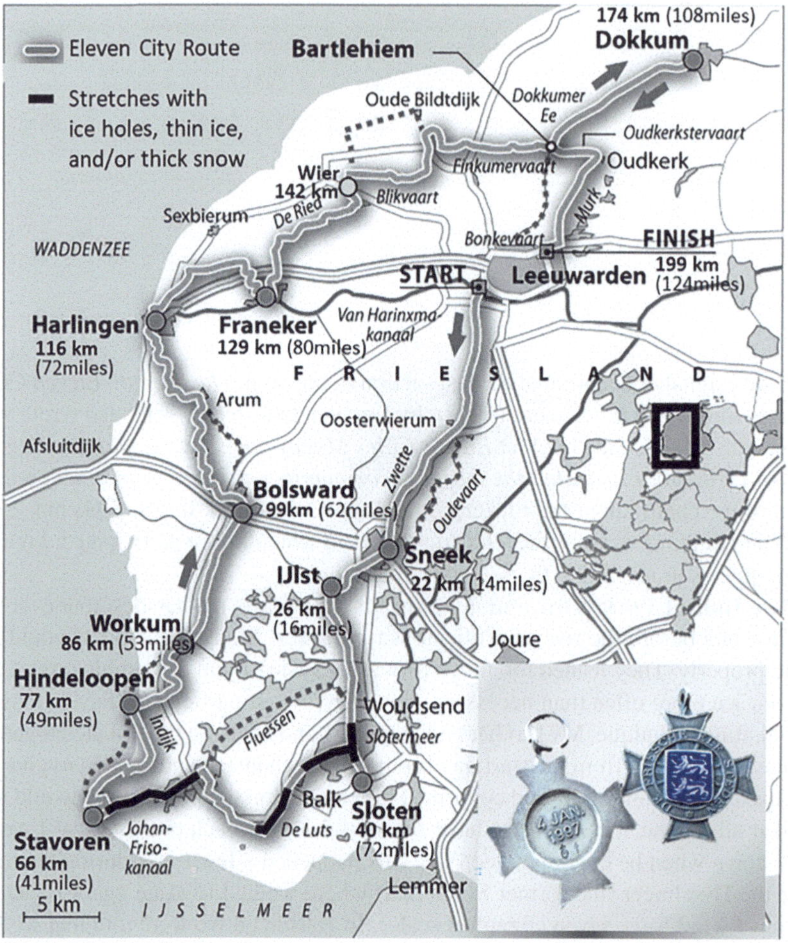

Eleven City Route

Stretches with ice holes, thin ice, and/or thick snow

174 km (108miles)
Dokkum

Bartlehiem

Oude Bildtdijk

Dokkumer Ee

Oudkerkstervaart

Finkumervaart

Oudkerk

Wier 142 km

Blikvaart

Sexbierum

De Ried

Bonkevaart

FINISH

WADDENZEE

START

Leeuwarden

199 km (124miles)

Harlingen 116 km (72miles)

Franeker 129 km (80miles)

Van Harinxma kanaal

F R I E S L A N D

Arum

Oosterwierum

Afsluitdijk

Zwette

Oudevaart

Bolsward 99km (62miles)

Sneek 22 km (14miles)

IJlst 26 km (16miles)

Joure

Workum 86 km (53miles)

Hindeloopen 77 km (49miles)

Indiik

Fluessen

Slotermeer

Woudsend

Balk

De Luts

Sloten 40 km (72miles)

Stavoren 66 km (41miles)

Johan-Friso-kanaal

Lemmer

5 km

I J S S E L M E E R

Fig. D.2 The Hell of '63

"The Hell of 1963"

I attempted in vain. But the assurance that nobody would come farther than where we were was enough reason for my brother to give up. Sixty-nine tour skaters—out of the 10,000 who started—ultimately finished. So, 58 tour skaters behind us must have continued after the police had left. When Wijnanda and I were discussing with Ir. Sipkema our proposal to give her the opportunity to write her master thesis about the management of the Eleven Cities Tour, I remarked that without the police interference, I would have finished at 10 o'clock. Ir. Sipkema resolutely responded: 'Impossible, you could not have finished at ten but you could at half past eleven', half an hour before the closing deadline of 12 o'clock.

Table D.1 Statistics of Eleven Cities tour race and tour participants.

Eleven Towns Tour Participation and Completion Statistics	The Tour Race Participants (Start 5:15am, Finish limit is 'Winner's finish time plus 1/5th of his/her race time')			The General Tour Participants		
Date	Number of skaters started	Number of skaters finished	Percentage completed the Tour	Number of skaters started	Number of skaters finished	Percentage completed the Tour
1 1909-01-02	23	7	30%	0	0	n.a.
2 1912-02-07	38	14	37%	22	4	18%
3 1917-01-27	42	9	21%	108	83	77%
4 1929-02-12	98	11	11%	206	103	50%
5 1933-12-16	173	57	33%	339	173	51%
6 1940-01-30	688	40	6%	2.716	27	1%
7 1941-02-06	600	65	11%	1.900	1.672	88%
8 1942-01-22	970	277	29%	3.862	3.669	95%
9 1947-02-08	277	39	14%	1.791	270	15%
10 1954-02-03	138	63	46%	2.597	2.143	83%
11 1956-02-14	259	109	42%	6.070	4.739	78%
12 **1963-01-18**	**568**	**57**	**10%**	**9.294**	**69**	**1%**
13 1985-02-21	227	224	99%	16.000	13.066	82%
14 1986-02-26	317	201	63%	16.999	14.788	87%
15 1997-01-04	301	122	41%	16.387	11.526	70%

Source: www.elfstedensite.nl/statistieken-elfstedentochten/

The Tour of 1985: A godsend for which my generation had to wait for 22 long years. A track with relatively thin ice, lots of 'Klunen' (walking over land to pass small stretches where the ice is too thin), and with the melting of ice during the relatively warm day. When I became aware of the melting ice, I skated as fast as I could to Stavoren. With flashbacks of 1963, I thought 'If they stop the tour, it should be behind me'.

The Tour of 1986: What I would consider to be ideal circumstances: beautiful ice and a friendly bit of sunshine.

The Tour of 1997: For me, and I suspect many others, I remember this tour as the 'head wind' tour. Strong head wind all the way from Stavoren to Dokkum, 108 km without interruption. The weather forecast of a decreasing wind in the afternoon never materialised. Also, the prediction of a bit of sunshine turned out to be false. Moreover, there had hardly been time for training. I only put on the skates four times before the tour and none of these trainings was over 50 km. In 1997, it took me 2.5 h longer than the previous two Tours in which I finished about half past 4 in the afternoon. The head wind and the limited training made it hard, but much was compensated by the wonderful quality of the ice and the heart-warming support from the people along the track. That support was present even in remote places 'yn 'e wrâld', a Frisian expression for 'in the middle of nowhere', and literally the endless flat green—now white—farm pastures that are so characteristic of the Frisian landscape.

What did not change? The outstanding organisation—an example I am convinced businesses should learn from—the enthusiasm of all people involved, but above all the hospitality of the family Vlas who provided me with lodging before and after the tour, like so many Frisian homes did for over 16,000 skaters. Their son graduated on an internship project in Taiwan. Later on, he married a Taiwan lady. He worked and lived in the Far East for over 30 years. My wife is Taiwanese by birth as well, but grew up in Brazil from the age of 14. This helped me to explain the attention the Eleven Cities Tour gets. 'It's like Carnival in Rio, but less predictable. You sometimes need to wait for many years'. She nodded with understanding and adds 'and it is a hell of a lot colder'.

At a quarter past 6 I am in the cage of the third starting group. Besides me is a woman who looks extraordinarily fresh. I can't stop myself from starting a conversation. It turns out she is from the province of Groningen. I asked, in view of our low participant number, if she had also participated in the Tour of '63? 'No, at that time I still laid in my maternity bed. I participated in '85 and '86. I then skated together with my husband'. To my question where he was now, she responded he had passed away 9 months ago. Being an active member of the Eleven Cities Tour Association, she had decided to skate now by herself. 'Not with your son?', I asked. 'No, the membership number had to be returned, nobody was able to prevent that'. I wished her well and reflected for some time on what she had told me. Wouldn't it be a good idea to allow to transfer the number to a family member? This would certainly accelerate participation by the younger generation. I would be a lot more prepared to release my number if I can transfer it to my son instead of letting it disappear in the crowd.

KONINKLIJKE VERENIGING
DE FRIESCHE ELF STEDEN

Official name and logo of the Royal Frisian Eleven Cities Association

White Veins

At 6.20 in the morning I depart with the light from only a sliver of a moon. It is enough though to see the cracks in the ice that run like white veins through the ice. The speed is high with the wind in the back. Within just 50 min I reach Sneek. Why some people are trying to go even faster than the wind I do not understand. It will be harsh enough after Stavoren.

A short train of seasoned marathon skaters passes me. They are casually chatting 'Hulzebosch doesn't know who Wiegel is. (Wiegel is the representative of the queen in the province of Friesland)'. The rest I cannot hear. A pity that I cannot maintain such speed anymore, shoots through my mind. I would like to follow the conversation a bit longer. My age is a significant difference across my Eleven Cities Tour experiences. In '63 I was 25, in the 'Sipkema' era in '85 I was in my late 40s, and now in my late 50s. A short acceleration to be able to follow an interesting conversation isn't possible anymore. The spending of energy has to be carefully managed. Only accept to be the lead skater when one has to. In '63 I didn't give a damn to be in the lead, more than enough energy would be left anyway. Now things are different. I skated for a while behind a skater from Groningen whose speed was about the same as mine. He complained that I had to contribute my share as lead skater. Fair enough. We did, though, compare our ages to determine who ought to lead somewhat more. He was 8 years younger than me, so we agreed he would lead a bit more.

Apparently, my age was also a worry for the person who had invited me to write this section of his book. In the evening before the tour, he gave me an anxious telephone call: What to do if I would give up somewhere during the tour? In utter disbelief I answered immediately 'That won't happen. I will think about how to distribute my energy, but quitting will never cross my mind'.

At 1 o'clock I enter Harlingen. Unbelievable, I count five bands playing wind instrument music! In particular the first two are playing very well. The first one plays typical marching band music; the second one played also Dixieland. Impressive, how they played in tune at such freezing temperatures! For a moment I am tempted to stop and listen for a while. I am quite well on schedule. I decide to keep going. It is still a long way to Dokkum, and it's better to keep some reserve. Also, I had promised the sport photographer to be in Wier by 3 in the afternoon, where I suspected to be a secret check point.

I indeed arrived in Wier at 3 o'clock, but a photographer was nowhere to be seen. I waited for 5 min. All right, then no photograph. The waiting made me realise that I was quite tired. I decided to take it easy even though it would mean skating in the dark somewhat longer. When you go full speed too early it can not only cause significant slowdowns later on, but can also lead to serious falls. I fell four times, three of which caused by another skater running in to me. In '63 I fell at least 30 times, all due to the terrible condition of the track.

Fig. D.3 Author's actual Eleven City Tour check point stamp card from the Tour in 1985

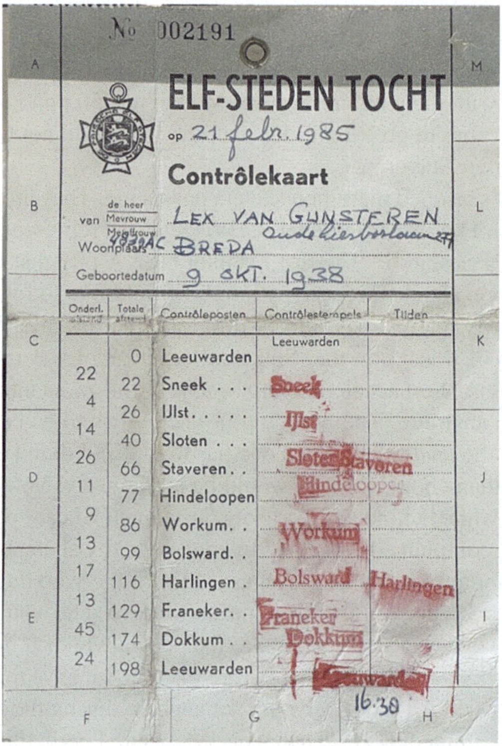

Drunkard

When I finally reach Dokkum, darkness is setting in. In the previous tour of 1986 I skated here in the full sun. Apparently too fast. We had to pass a stretch by 'klunen'. I was so exhausted that I swaggered like a drunk while walking on skates on the 'kluun' track. A teenage girl crawled under the fence and asked 'Shall I support you sir?' Wonderful! I only had to place one foot in front of the other. No need to keep direction, since she took good care of steering and balance. Skaters who were already going back gave me envious looks. What a morale booster it was, my dip evaporated instantly. How would she be now, in 1997, an attractive woman in her early 30s? This time, no 'klunen' needed. Only a crowd of enthusiastic people cheering on the skaters who had just overcome over hundred kilometres of hard head wind. Heart-warming, although very different from last time.

I take some food and start my way back from Dokkum. Like the stretch at the start, the wind is strong in the back. But visibility is even less. The morning's sliver of moon is absent. When I slide my ski glasses up to improve vision, my eyes start tearing and freezing. I put my ski glasses on again and keep my direction by sensing the ice surface and taking my time. After the second time passing Bartlehiem, there is a short stretch with head wind. I do not join a group. When visibility is poor I rather have nobody in front of me. After the secret check point in Oudekerk, the wind is in the back again. A few kilometres before the finish, someone tries to marginally improve his time by skating faster than the wind. Just before he would pass me he falls and glides over the ice as a curling, exactly into my track. I feel my legs being wiped away and fall painfully on my left side. Could it happen to be forced to quit 3 km from the finish? In my anger, I tell the man in no unclear terms what I think of him. A moment later, I regret my outburst because such is not proper in an Eleven Cities Tour.

The Finale

At 7 o'clock all is finished. I step onto the bus which is part of the excellent organisation. I sink into the front seat. When I tell my neighbour that he is sitting in a seat reserved for handicapped people, he says 'That's exactly how I feel'. After handing in my card at the Frieslandhal, I walk on my clog shoes back to my lodging address. Apparently, my walking was not very smooth anymore, for I was offered a ride by somebody who had finished ahead of me, 'I know how you feel, that is how I walked an hour ago'. He drops me off in front of the house. Mr. Vlas opens the door. He congratulates me and asks what I want, a beer or kale with sausage. 'A bath or a shower' I answer. Before going into the bath, I ask how the other skaters have performed. It turns out that I am the second one who had finished. In the bath, my muscles enjoying the hot water, I realise: I am not the fastest participant anymore in the house of the family Vlas. The times of finishing within a challenging time are over. I have entered the era of trying to finish without too many bruises and within the time limit. Like I heard Jeen van den Berg (winner of the Eleven City Tour in 1954, third place in '63, and 7-time finisher) once comment: 'Of course, you want to show that you can still do it'. I couldn't agree more.

Eleven Cities Tour monument, near Giekerk—it is the last bridge the tour skaters pass under with one last kilometre left to the finish in the capital of Friesland, Leeuwarden. Between 2001 and 2008 the bridge was tiled with over 4000 blue-on-white glazed tiles with the images of over 7000 skaters with their names and year(s) they finished the tour. Together they form a mosaic portraying the silhouette of skaters. The south side of the bridge is still bare, awaiting the tiles of future skaters. The tile in the center of the figure is one of the 4,000 tiles, and shows Van Gunsteren resting in the bath of family Vlas after finishing the Tour in '97. To the left are the original skates he learned to skate on during WWII.

Bibliography

Abrami, Regina M., Kirby, William C., and McFarlan, F. Warren, *"Why China Can't Innovate"*, Harvard Business Review, March 2014.

Ackoff, Russell L., *Creating the Corporate Future*, John Wiley & Sons, 1981.

_____, *The Democratic Corporation,* Oxford University Press, 1994.

_____, Ackoff's Best, His Classic Writings on Management, John Wiley & Sons, 1999.

Binnekamp, Ruud, *Preference based Design in Architecture,* PhD thesis Delft University of Technology, 2005.

Binnekamp, R., van Gunsteren, L.A., van Loon, P.P., *Open Design, a Stakeholder-oriented Approach in Architecture, Urban Planning and Project Management,* IOS Press, Amsterdam, 2006.

Breitinger, Jan C., Dierks, Benjamin, Rausch, Thomas, *"World class patents in cutting-edge technologies"*, BertelsmannStiftung, 06 March 2020.

Black, J. Stewart, Morrison, Allen J., "The Strategic Challenges of Decoupling", *HBR Magazine*, May-June 2021.

Bult, Frans, Straatman, John, "Opkomst en groei van een Noord Brabantse scheepsschroevenfabrikant", Histordia Bokhoven, 2011.

Christensen, Clayton M. *The Innovator's Dilemma.* Harvard Business Review Press, 2013.

Cusumano, Michael A., Yoffie, David B., and Gawer, Annabelle, "The Future of Platforms", *MIT Sloan Management Review*, Spring 2020.

_____, *The Business of Platforms,* Harper Collins, 2018.

Dobbs, Richard., Manyika, James., Woetzel Jonathan. "The China Effect on Global Innovation", *McKinsey Global Institute*, 2015.

Drucker, Peter F., *Innovation and Entrepreneurship, Practice and Principles,* Heinemann, London, 1985.

Dychtwald, Zak, "China's New Innovation Advantage", *HBR Magazine*, May-June 2021.

Gawer, Annabelle, and Michael A. Cusumano. "Industry Platforms and Ecosystem Innovation." Journal of Product Innovation Management 31, no. 3, (September 4, 2013): 417–433.

Grove, Andy S. *Only the Paranoid Survive,* Profile Books Ltd, 1997.

van Gunsteren, Frans F., *China's Need for Small Northern European Friends,* Parthenon Publishing House, 2011.

van Gunsteren, Herman R., *The Quest for Control, a critique of the rational-central-rule approach in public affairs,* John Wiley & Sons Ltd, 1976.

van Gunsteren, Lex A., "Ring Propellers and their combination with a stator", *Marine Technology,* October 1970.

© The Author(s), under exclusive license to Springer Nature Switzerland AG 2022

L. A. van Gunsteren, A. G. Vlas, *The License Giver Business Concept of Technological Innovation*, Future of Business and Finance,

https://doi.org/10.1007/978-3-030-91123-2

_____. "Ring propellers", *Transactions of the Institute of Marine Engineers,* Vol. 83, Part 2, 1971.

_____. "Slotted Nozzles", Paper presented at the North American Tug Convention, Vancouver, Canada, April 30th-May 3rd, 1973. Included as Chapter 7 of van Gunsteren's PhD thesis, Delft University of Technology, December 12, 1973a.

_____. *A Contribution to the Solution of some Specific Ship Propulsion Problems, A Reappraisal of Momentum Theory,* Ph.D. Thesis, Delft University of Technology, 1973b.

_____. and Gibson, Ivan S., "Optimisation of Nozzle-Propeller-Rudder Configurations for High Powered Large Ships*", Symposium on high powered propulsion of large ships, NSMB,* Wageningen, December 10-13, 1974. Also, *International Shipbuilding Progress,* Vol. 22, No.246, February 1975.

_____. *Information Technology, A Managerial Perspective-* In: *New Technology as Organisational Innovation. The Development and Diffusion of Microelectronics,* edited by Johannes M. Pennings and Arend Buitendam, Ballinger Publishing Company. pp.277–289. 1987.

_____ and Willemse, Wijnanda J., Elfsteden Management, Het management van de Elstedentocht en wat het bedrijfsleven ervan kan leren, Uitgeverij van Wijnen, Franeker, 1988.

_____ and Gelling, Jaap L., "Flapped Nozzles", *Seventh Lips Propeller Symposium,* September 20th-22nd, 1989.

_____ and Gelling, Jaap L., "G&G Flapped Nozzles: Past and Current Developments", *Scandinavian & European Shipping Review,* Winter, 1991.

_____ "Crisis Management", Financial Times Handbook of Management, pp 812–829, Pitman Publishing, 1995.

_____. De vijftiende Elfsredentocht 1997 Heldenstrijd in ijzige wind, Tirion-Baarn, 1997.

_____. "Architectural Design: Skill, Science or Religion", R*esearch by Design, International Conference,* Delft, 1–3 November 2000.

_____ and van Loon, P.P., *Open Design, A Collaborative Approach to Architecture,* Eburon Publishers, 2000.

_____. *Management of Industrial R&D, A Viewpoint from Practice.* Third Revised Edition, Eburon Publishers, 2003a.

_____. Open Design Methodology in Governmental decision Making: The Case of Schiphol Airport Amsterdam, *Mathematical and Computer Modelling,* 2003b.

_____. *On Innovation, Essays by Lex van Gunsteren on the occasion of his retirement from Delft University of Technology,* including a Liber Amicorum, Eburon Publishers, 2004.

_____. *Stakeholder-oriented Project Management,* IOS Press, Amsterdam, 2011.

_____. *Leading Professionals the Natural Way,* Parthenon Publishing House, 2012.

_____. *Quality in Design and Execution of Engineering Practice.* IOS Press, Amsterdam, 2013.

_____. "Continuous Adjustments and the Reality Test in Managing Complex Projects", *Journal of Modern Project Management,* May August 2020.

Gupta, Anil K, and Wang, Hayan, " How China's Government Helps - and Hinders - Innovation", *Harvard Business Review,* November 2016.

D'Hooghe, Ingrid , *The 1991/1992 Dutch debate on the sale of submarines to Taiwan,* March 1 1992.

Huibregtsen, Micky. *Management made easy, Ideas of a former McKinsey Partner.* Dexter, Bakas Books, 2017.

Isaacson, Walter, *Steve Jobs,* Simon and Schuster, 2011.

Mansfield, E. *Industrial Research and Technological Innovation.* Norton New York, 1968.

Mitter, Rana, Johnson, Elsbeth, "What the West Gets Wrong About China", HBR Magazine, May-June 2021.

Mintzberg, Henry, *The structuring of Organizations,* Prentice Hall Inc., 1979.

Moss Kanter, Rosabeth, *The Change Masters,* Simon & Schuster Inc., 1983.

Mulder, M. "The Logic of power and the urgency of good leadership in our Atlantic society", 2010, Free to download from www.maukmulder.nl. http://www.maukmulder.nl/download.php

Overdiek, Markus and Coka, Daniela Arregui, "Industrial Policy – Lessons from China", *Global Economic Dynamics (GED),* February 2020.

Pisano, Gary P. "The hard truth of innovating cultures", *Harvard Business Review,* January-February 2019.

Podolny, Joel M and, Hansen, Morten T, "How Apple Is Organized for Innovation, it's about experts leading experts", *Harvard Business Review*, November-December, 2020.

Porter, Michael E. "Clusters and the New Economics of Competition", *Harvard Business Review*, November-December 1998.

Prud'homme, Dan, Zedtwitz, Max von, "The Changing Face of Innovation in China", *MIT Sloan Management Review*, 12 June 2018.

Schön, Donald A. *Beyond the Stable State.* New York: Random House, 1971.

_____. "The Fear of Innovation". Chapter 8 of *"The R&D Game"*, Fourth printing, MIT Press, 1977.

Simon, M. *De Strategische Functie Typologie,* PhD Thesis, Leiden University, 1989.

Webb, Michael, Short, Nick, Bloom, Nicholas, Lerner, Josh, "Some Facts of High-Tech Patenting", *Harvard Business School Working Paper* 19-014, July 2018.

Nomenclature

c Chord

C_T Thrust coefficient defined as: $T = \dfrac{T}{\dfrac{1}{2}\rho V_\mathrm{A}^2 \dfrac{\pi}{4} D^2}$

D Propeller diameter

$\bar{D}$ Outside diameter of propulsion device

J Advance ratio defined as: $J = \dfrac{V_\mathrm{A}}{nD}$

n Revolutions per unit of time

η_0 Open water efficiency

t Thickness

V_A Speed of advance

α Angle of attack

β Drift angle $= \delta - \alpha$

δ Rudder angle

ρ Density

© The Author(s), under exclusive license to Springer Nature Switzerland AG 2022
L. A. van Gunsteren, A. G. Vlas, *The License Giver Business Concept of Technological Innovation*, Future of Business and Finance,
https://doi.org/10.1007/978-3-030-91123-2

Author Index

Subject Index

A
Acer, 63
Adhocracy, 53, 55, 56
Airbus, 61
Airport island, 67
Alignment, 60–65
ARM, 66
Autocratic, 73

B
Basic research, 75
Best-management practices, 110
Boeing, 3, 61
Breakthrough innovation, 75, 113–119

C
Cartel, 43, 44, 50, 116, 118
Capacity, 5, 7, 12, 16, 60, 69, 85, 87, 88, 92–94, 103, 105
Challenges, 2, 51–52, 54, 109
Chief Executive Officer (CEO), 6, 51, 55, 69, 106, 109
Chief Financial Officer (CFO), 54, 55
China, 30, 41, 58, 70, 71, 73–77, 105, 109
Clusters of competence, 67, 75
Commercialisation, 2, 14, 17, 86, 88, 102
Compaq, 6, 63
Compliance, 13, 28, 105
Compromise, 67, 111
Computer, xiii, 1, 15, 16, 18, 19, 27, 37, 47, 49, 59, 62–66, 80, 81
Conformance, 12, 13
Consultancy, 6, 7, 15, 69
Consultants, 3, 5, 14, 15, 27, 28, 48, 69, 85, 89, 102, 103

Cosmetic quality, 10, 11, 109
COVID-19, xiv, 1, 37
Creativity, 2, 17, 38, 74, 75
Creator, 14
Crucial quality, 10, 108, 109
Customers, 16, 19, 29, 50, 88, 102, 103, 105

D
DEC, 62
Decentralisation, 54
Decline, 51–58
Decoupling, 76, 77
Deductive thinking, 75
Dell, 6, 7, 63
Democracy, 73, 76
Design leadership, 2, 5, 14, 18, 50, 63, 64, 75, 89, 92
Design quality, 3, 8–13, 55, 56
Development, xiii, 2, 19, 25, 29, 32, 37, 41–43, 47, 49–52, 59–61, 73–76, 79, 80, 86–90, 93, 95, 97, 98, 101, 117, 119
Diffusion of innovation, 1, 13, 14, 83
Dilemma, 5, 27, 28, 66, 67, 85, 89, 100
Discoveries, 13, 14
Divisions, 46, 68, 86, 87, 92, 100–102
Doctrine, 52, 55, 56

E
Economy, 15, 35, 72–74, 76, 107
Economy-of-scale, 60, 62, 64, 66
Education, 16, 70
Effectiveness, 4, 5, 14, 19
Efficiency, 5, 14, 19, 26, 27, 64
Eleven Cities, 22
Entrepreneurship, 70, 75

© The Author(s), under exclusive license to Springer Nature Switzerland AG 2022
L. A. van Gunsteren, A. G. Vlas, *The License Giver Business Concept of Technological Innovation*, Future of Business and Finance,
https://doi.org/10.1007/978-3-030-91123-2

The manufacturer's authorised representative in the EU is Springer

Nature Customer Service Centre GmbH, Europaplatz 3, 69115 Heidelberg,

Germany. If you have any concerns regarding our products, please

contact ProductSafety@springernature.com

Printed and bound by CPI Group (UK) Ltd, Croydon, CR0 4YY

29/04/2026

02099522-0009